华章IT | HZBOOKS | Information Technology

智能系统与技术丛书

Neural Network Programming with TensorFlow

TensorFlow
神经网络编程

［印度］曼普里特·辛格·古特（Manpreet Singh Ghotra）著
　　　　拉蒂普·杜瓦（Rajdeep Dua）

马恩驰 陆健 译

机械工业出版社
China Machine Press

图书在版编目（CIP）数据

TensorFlow 神经网络编程 /（印）曼普里特·辛格·古特（Manpreet Singh Ghotra），（印）拉蒂普·杜瓦（Rajdeep Dua）著；马恩驰，陆健译 . —北京：机械工业出版社，2018.10
（智能系统与技术丛书）

书名原文：Neural Network Programming with TensorFlow

ISBN 978-7-111-61178-3

I. T⋯ Ⅱ.①曼⋯ ②拉⋯ ③马⋯ ④陆⋯ Ⅲ. 人工神经网络－程序设计 Ⅳ.① TP183 ② TP311.1

中国版本图书馆 CIP 数据核字（2018）第 234716 号

本书版权登记号：图字　01-2018-2519

Manpreet Singh Ghotra, Rajdeep Dua: *Neural Network Programming with TensorFlow* (ISBN: 978-1-78839-039-2).

Copyright © 2017 Packt Publishing. First published in the English language under the title "Neural Network Programming with TensorFlow".

All rights reserved.

Chinese simplified language edition published by China Machine Press.

Copyright © 2018 by China Machine Press.

本书中文简体字版由 Packt Publishing 授权机械工业出版社独家出版。未经出版者书面许可，不得以任何方式复制或抄袭本书内容。

TensorFlow 神经网络编程

出版发行：机械工业出版社（北京市西城区百万庄大街 22 号　邮政编码：100037）	
责任编辑：唐晓琳	责任校对：殷　虹
印　　刷：北京诚信伟业印刷有限公司	版　次：2018 年 11 月第 1 版第 1 次印刷
开　　本：186mm×240mm　1/16	印　张：14.75
书　　号：ISBN 978-7-111-61178-3	定　价：69.00 元

凡购本书，如有缺页、倒页、脱页，由本社发行部调换
客服热线：（010）88379426　88361066　　　投稿热线：（010）88379604
购书热线：（010）68326294　88379649　68995259　　读者信箱：hzit@hzbook.com

版权所有·侵权必究
封底无防伪标均为盗版
本书法律顾问：北京大成律师事务所　韩光 / 邹晓东

THE TRANSLATOR'S WORDS

译 者 序

以深度学习为代表的人工智能在计算机视觉、语音识别、自然语言处理方面已经取得了突破性的进展。越来越多的公司更倾向于使用深度学习模型来解决业务场景中的问题。在众多的计算框架中，作为 Google 最新开源的深度学习计算框架，TensorFlow 支持包括 CNN、RNN、LSTM 等多种神经网络模型。由于其强大丰富的功能、高度灵活的 API，目前 TensorFlow 受到了业界的高度关注。

TensorFlow 作为主流的深度学习框架仍在不断地演进。北京时间 2018 年 3 月 31 日，美国加利福尼亚州山景城的计算机历史博物馆举办了第二届 TensorFlow 开发者峰会，吸引了超过 500 名 TensorFlow 用户亲临现场，数万名来自世界各国的观众观看了本次峰会的直播。活动当天，TensorFlow 团队和邀请嘉宾举办了多场技术讲座。编程语言方面，TensorFlow 会支持越来越多的 ML 算法，对接以 js 和 swift 为代表的越来越多的编程语言。数据 IO 方面，输入数据的处理非常重要，良好的深度学习框架不仅需要支持不同类型的数据源以及数据格式，同时还要保证高吞吐率，以满足日益发展的硬件需求。本次大会引入了 tf.data API。tf.data 采用了标准数据库领域里的 ETL 概念，将主要的接口封装起来。并且可以通过 chaining 的方式组合。加载图片数据的速度可达 13k/s。上层 API 上，大会还是推广 Estimator 的使用，Estimator 在特征工程上提供了更便捷的上层 API，可使特征值分箱、特征交叉、特征嵌入更便捷。值得一提的是，在浏览器上支持深度学习意义重大，因为浏览器可以有更好的可视化能力，在模型调参和增加结果的可解释性方面大有裨益，比如 TensorFlow Playground 就是成功的案例。

在商业领域应用方面，国内各大公司都在广泛使用 TensorFlow，并在分布式训练、模型压缩与加速、在线预测方面进行了优化。在场景应用方面，我们团队开发的个性化表征学习和深度语义模型已全量上线，并取得了不错的 abtest 效果，训练框架均基于 TensorFlow。其中，个性化表征学习主要应用于搜索排序，通过视频的文本信息和统计特征编码对视频进行表达，并通过视频观看的 i2i 和 u2i 分别在视频观看序列上对相关视频和用户视频观看兴趣做个性化表达。最后，基于曝光点击行为数据处理得到标签以进行训练，线上效果证明，用户在头部宽泛意图下对视频内容有较大的个性化需求。相关性计算的深度语义模型，通过对视频的文本信息进行编码，基于人工标注的标签数据进行训练。深度语义模型在尾部搜索意图下的视频结果有明显的体验提升。

本书简明扼要地介绍了如何基于 TensorFlow 实现神经网络，并在代码实现方面做了详尽的说明。全书由我和陆健翻译完成，在繁重的工作之余挤出时间完成翻译需要极大的毅力。其中，我参与翻译了第 1、3、5、7、9 章，陆健参与翻译了第 2、4、6、8、10 章。由于水平有限，书中可能会有翻译不妥之处，读者在阅读过程中若发现不妥之处，欢迎发邮件至 maec1208@gmail.com。

感谢华章公司的王春华编辑为本书翻译及出版过程中所做的努力，感谢我的爱人和孩子在翻译过程中给予的支持，你们是我坚持奋斗的最大动力。

马恩驰

2018 年 9 月于杭州

ABOUT THE AUTHORS
作者简介

Manpreet Singh Ghotra 在企业和大数据软件方面拥有超过 15 年的软件开发经验。目前，他正致力于开发一个机器学习平台 /API，该平台主要使用诸如 TensorFlow、Keras、Apache Spark 和 PredictionIO 等开源库和框架进行开发。Manpreet Singh Ghotra 在各种机器学习应用场景上有丰富的经验，其中包括情感分析、垃圾邮件检测、图像调整和异常检测。他是世界上最大在线零售商之一机器学习组的成员，主要工作是使用 R 和 Apache Mahout 做运输时间优化。他拥有机器学习方面的研究生学位，为机器学习社区工作并贡献卓越。

Manpreet Singh Ghotra 的 GitHub 地址为 https://github.com/badlogicmanpreet。你也可以在 LinkedIn 上的 https://in.linkedin.com/in/msghotra 找到他。

> 本书献给我的父母 Amrik Singh 和 Nirmal Kaur。感谢他们在我写作本书时，对我的支持。

Rajdeep Dua 在云计算、大数据和机器学习领域拥有超过 18 年经验。他曾供职于 Google 的大数据工具 BigQuery 团队，也曾在 VMware 旗下的 Greenplum 大数据平台推广团队担任开发人员。他还与一个团队紧密合作并致力于把 Spark 作为一个功能模块移植到 VMware 的公有云和私有云上。Rajdeep Dua 曾在印度一些声望较高的科技学校（海得拉巴国际教育学院、印度国际学校、德里国际教育研究所和浦那工程学院）负责 Spark 和大数据的教学工作。

目前，他在印度 Salesforce 公司领导开发者关系团队。他就 Salesforce 和 Apache Prediction IO 的机器学习库 Einstein 发布了很多博客和动手实验。

Rajdeep Dua 在 cluddatalab 上发布了很多关于大数据和 Spark 的课程。他还在海德拉巴的 W3C 会议上发布了 BigQuery 和 Google App Engine。他曾领导谷歌、VMware 和微软的开发者关系团队，并在云计算的数百个会议上发表过演讲。

Rajdeep Dua 对开源社区的贡献主要针对 Docker、Kubernetes、Android、OpenStack、预测 IO 和 cloudfoundry 项目。

    本书献给我的父母。感谢我的妻子 Manjula 以及儿子 Navtej 和 Kairav 在我花时间写作本书的时候陪伴我。

审校者简介

Giancarlo Zaccone 拥有十多年管理科学和工业研究项目的经验。他曾在美国国家研究委员会 C.N.R 担任研究员，在那里他参与了与并行数值计算和科学可视化相关的项目。

目前，他是一家咨询公司的高级软件工程师，负责开发和测试太空与国防应用软件系统。

Giancarlo 拥有那不勒斯费德里克二世的物理学硕士学位和罗马大学的二级科学计算硕士学位。

他是以下 Packt 出版书籍的作者：《Python Parallel Programming Cookbook》《Getting Started with TensorFlow》和《Deep Learning with TensorFlow》。

可以通过 https://it.linkedin.com/in/giancarlozaccone 联系他。

PREFACE

前　　言

如果你意识到到处都有围绕机器学习、人工智能或深度学习等术语的讨论，你可能会知道什么是神经网络。想知道如何利用它们有效地解决复杂的计算问题，或者怎样训练有效的神经网络？本书将教你所有这些以及更多的事情。

首先快速浏览流行的 TensorFlow 库，并了解如何用它训练不同的神经网络。之后你将深入了解神经网络的基础知识和它背后的数学原理，以及为什么选择 TensorFlow 训练神经网络。然后，你将实现一个简单的前馈神经网络。接下来，你将掌握使用 TensorFlow 进行神经网络优化的技术和算法。更进一步，你将学习如何实现一些更复杂的神经网络，如卷积神经网络（CNN）、递归神经网络（RNN）和深度信念网络（DBN）。在学习本书的过程中，为了使你对神经网络编程有深切的理解，将在真实世界的数据集上训练模型。你还将训练生成模型，并学习自编码器的应用。

在本书的最后，你将对如何利用 TensorFlow 的强大功能来训练各种复杂的神经网络有一个正确的理解，而不会有任何困惑。

本书内容

第 1 章介绍神经网络中基本的代数知识、概率论和优化方法。

第 2 章介绍感知机、神经元和前馈神经网络的基础知识。你还将学习各种模型学习的技巧，并主要学习称为反向传播的核心学习算法。

第 3 章介绍对神经网络学习至关重要的优化方法。

第 4 章详细讨论 CNN 算法。CNN 及其在不同数据类型中的应用也包含在该章中。

第 5 章详细介绍 RNN 算法。RNN 及其在不同数据类型中的应用也包括在该章中。

第 6 章介绍生成模型的基础知识以及不同的生成模型。

第 7 章包括深度信念网络的基础知识、它们与传统神经网络的区别以及它们的实现。

第 8 章介绍最近处在生成模型前沿的自编码器。

第 9 章讨论深度学习当前和未来的具体研究内容,并包括一个参考文献。

第 10 章讨论 TensorFlow 的环境配置、TensorFlow 与 Numpy 的比较以及自动微分的概念。

准备工作

本书将指导你完成所有书中例子所需工具的安装:

- Python3.4 或更高版本。
- TensorFlow 1.4 或更高版本。

读者对象

本书适合希望使用神经网络的拥有统计背景的开发人员。虽然我们将使用 TensorFlow 作为神经网络的基础库,但本书可以作为从深度学习数学理论向实际应用转化的通用资源。如果你对 TensorFlow 和 Python 有一些了解并希望知道比 API 更底层的一些情况,本书会很适合你。

本书约定

表示警告或重要的注意事项。

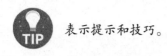 表示提示和技巧。

下载示例代码及彩色图片

本书的示例代码及所有截图和样图,可以从 http://www.packtpub.com 通过个人账号下载,也可以访问华章图书官网 http://www.hzbook.com,通过注册并登录个人账号下载。

目 录

译者序
作者简介
审校者简介
前言

第1章 神经网络的数学原理 ·················· 1
1.1 理解线性代数 ························· 1
1.1.1 环境设置 ······················ 2
1.1.2 线性代数的数据结构 ········ 3
1.1.3 线性代数运算 ················ 4
1.1.4 求解线性方程 ················ 9
1.1.5 奇异值分解 ·················· 11
1.1.6 特征值分解 ·················· 14
1.1.7 主成分分析 ·················· 14
1.2 微积分 ································ 15
1.2.1 梯度 ··························· 16
1.2.2 Hessian 矩阵 ················· 23
1.2.3 行列式 ························ 24
1.3 最优化 ······························ 25
1.4 总结 ·································· 28

第2章 深度前馈神经网络 ·················· 29
2.1 定义前馈神经网络 ················· 29
2.2 理解反向传播 ······················ 30
2.3 在 TensorFlow 中实现前馈神经网络 ······························ 31
2.4 分析 Iris 数据集 ···················· 34
2.5 使用前馈网络进行图像分类 ····· 40
2.6 总结 ·································· 54

第3章 神经网络的优化 ····················· 55
3.1 什么是优化 ························· 55
3.2 优化器的类型 ······················ 56
3.3 梯度下降 ···························· 57
3.3.1 梯度下降的变体 ············· 58
3.3.2 优化梯度下降的算法 ······· 59
3.4 优化器的选择 ······················ 61
3.5 总结 ·································· 64

第4章 卷积神经网络 ························ 65
4.1 卷积神经网络概述和直观理解 ·· 66
4.1.1 单个卷积层的计算 ·········· 66
4.1.2 TensorFlow 中的 CNN ······· 70

4.2 卷积操作 ································· 72
   4.2.1 对图像进行卷积 ············· 73
   4.2.2 步长 ································· 75
4.3 池化 ·· 76
   4.3.1 最大池化 ························· 77
   4.3.2 示例代码 ························· 78
4.4 使用卷积网络进行图像分类 ······ 80
4.5 总结 ·· 102

第5章 递归神经网络 ················ 103

5.1 递归神经网络介绍 ················ 103
   5.1.1 RNN 实现 ····················· 105
   5.1.2 TensorFlow RNN 实现 ····· 110
5.2 长短期记忆网络简介 ············ 114
   5.2.1 LSTM 的生命周期 ········· 115
   5.2.2 LSTM 实现 ···················· 117
5.3 情感分析 ································ 122
   5.3.1 词嵌入 ··························· 122
   5.3.2 使用 RNN 进行情感分析 ···························· 128
5.4 总结 ·· 134

第6章 生成模型 ·························· 135

6.1 生成模型简介 ························ 135
   6.1.1 判别模型对生成模型 ····· 136
   6.1.2 生成模型的类型 ············· 137
6.2 GAN ·· 140
   6.2.1 GAN 示例 ······················ 141
   6.2.2 GAN 的种类 ·················· 150
6.3 总结 ·· 152

第7章 深度信念网络 ·················· 153

7.1 理解深度信念网络 ················ 154
7.2 训练模型 ································ 161
7.3 标签预测 ································ 162
7.4 探索模型的准确度 ················ 162
7.5 DBN 在 MNIST 数据集上的应用 ··· 163
   7.5.1 加载数据集 ····················· 163
   7.5.2 具有 256 个神经元的 RBM 层的 DBN 的输入参数 ······· 163
   7.5.3 具有 256 个神经元的 RBM 层的 DBN 的输出 ··············· 165
7.6 DBN 中 RBM 层的神经元数量的影响 ·· 165
   7.6.1 具有 512 个神经元的 RBM 层 ································ 165
   7.6.2 具有 128 个神经元的 RBM 层 ································ 166
   7.6.3 准确度指标对比 ············· 166
7.7 具有两个 RBM 层的 DBN ······· 167
7.8 用 DBN 对 NotMNIST 数据集进行分类 ································ 169
7.9 总结 ·· 172

第8章 自编码器 ·························· 173

8.1 自编码算法 ···························· 174
8.2 欠完备自编码器 ···················· 175
8.3 数据集 ···································· 175
8.4 基本自编码器 ························ 177
   8.4.1 自编码器的初始化 ········· 177
   8.4.2 AutoEncoder 类 ·············· 178

8.4.3　应用于 MNIST 数据集的
　　　　　基本自编码器……………180
　　8.4.4　基本自编码器的完整代码…184
　　8.4.5　基本自编码器小结………186
8.5　加性高斯噪声自编码器…………186
　　8.5.1　自编码器类………………187
　　8.5.2　应用于 MNIST 数据集的
　　　　　加性高斯自编码器………188
　　8.5.3　绘制重建的图像…………191
　　8.5.4　加性高斯自编码器的完整
　　　　　代码………………………192
　　8.5.5　比较基本自编码器和加性
　　　　　高斯噪声自编码器………193
　　8.5.6　加性高斯噪声自编码器
　　　　　小结………………………194
8.6　稀疏自编码器……………………194
　　8.6.1　KL 散度……………………194
　　8.6.2　稀疏自编码器的完整
　　　　　代码………………………196
　　8.6.3　应用于 MNIST 数据集的
　　　　　稀疏自编码器……………198
　　8.6.4　比较稀疏自编码器和加性
　　　　　高斯噪声自编码器………200
8.7　总结………………………………200

第 9 章　神经网络研究……………201
9.1　神经网络中避免过拟合…………201
　　9.1.1　过拟合问题阐述…………201

9.1.2　过拟合解决方案…………202
9.1.3　影响效果…………………203
9.2　使用神经网络进行大规模
　　视频处理……………………………204
　　9.2.1　分辨率改进方案…………204
　　9.2.2　特征直方图基线…………205
　　9.2.3　定量结果…………………205
9.3　使用双分支互向神经网络进行
　　命名实体识别………………………206
　　9.3.1　命名实体识别的例子……206
　　9.3.2　定义 Twinet………………207
　　9.3.3　结果………………………208
9.4　双向递归神经网络………………208
9.5　总结………………………………209

第 10 章　开始使用 TensorFlow……211
10.1　环境搭建…………………………211
10.2　比较 TensorFlow 和 Numpy……212
10.3　计算图……………………………213
　　10.3.1　图…………………………213
　　10.3.2　会话对象…………………214
　　10.3.3　变量………………………215
　　10.3.4　域…………………………216
　　10.3.5　数据输入…………………217
　　10.3.6　占位符和输入字典………217
10.4　自动微分…………………………218
10.5　TensorBoard………………………219

CHAPTER 1

第 1 章

神经网络的数学原理

神经网络使用者需要对神经网络的概念、算法和背后的数学知识有一个良好的理解。良好的数学直觉和多种技术的理解对于掌握算法的内在机制和获得良好模型效果是必需的。理解这些技术所需的数学知识和数学水平涉及多个方面,这也取决于你的兴趣。在本章中,你将通过了解用于解决复杂计算问题的数学知识来学习神经网络。本章内容涵盖了线性代数、微积分和神经网络优化的基础知识。

本章的主要目的是为接下来的章节建立数学基础。

本章将介绍以下内容:

- 线性代数。
- 微积分。
- 最优化。

1.1 理解线性代数

线性代数是数学的一个重要分支。理解线性代数对于**深度学习**(即神经网络)至关重要。在本章中,我们将学习线性代数的关键和基础知识点。线性代数主要处理线性方程组。我们从使用矩阵和向量开始,而非标量。使用线性代数,我们能在深度学习中描述复杂的操作。

1.1.1 环境设置

在进入数学及其性质的领域之前,构建开发环境必不可少,因为它将为我们提供环境来实践我们所学的概念知识。这意味着我们需要安装编译器及其依赖和 IDE(集成开发环境)来运行代码。

在 PyCharm 中设置 Python 环境

我们最好使用像 PyCharm 这样的 IDE 来编辑 Python 代码,因为它提供了开发工具和内置的编码帮助。代码检查使代码开发和调试更加简单快速,确保你专注于学习神经网络中的数学知识这个终极目标。

以下步骤将展示如何在 PyCharm 中设置本地 Python 环境:

1)请先到 Preferences 中确认你已经安装了 TensorFlow 库,如果没有安装,请参考 `https://www.tensorflow.org/install/` 上的说明安装 TensorFlow:

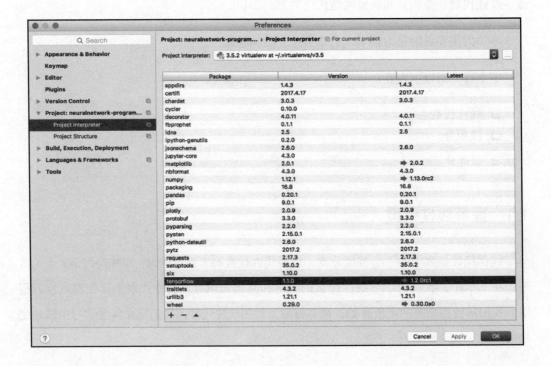

2）选择 TensorFlow 的默认选项然后点击 OK。

3）最后右击源文件点击 Run 'matrices':

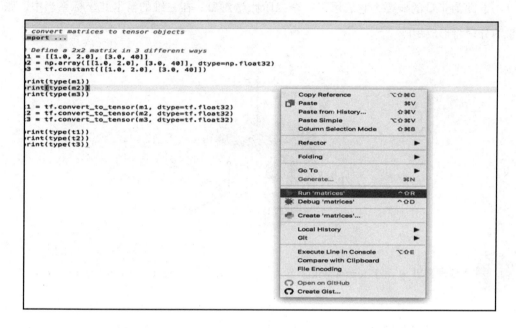

1.1.2 线性代数的数据结构

下面将阐述线性代数的基础数据结构。

1. 标量、向量和矩阵

标量、向量和矩阵是数学中的基本对象，基本定义如下：

- 标量由一个单独的数或数值（称为**量级**）来表示。
- 向量是有序排列的一列数。由唯一的索引来标识每个单独的数。向量表示空间中的一个点，向量中的每个元素给出了不同轴下的坐标。
- 矩阵是一个二维数组，矩阵中的每个数通过索引（i, j）来标识。

2. 张量

具有可变轴（即多维）的数组称为**张量**（tensor）。例如，对于三个轴，我们需要使用

索引 (i,j,k) 来确定每个独立的数。

下图为张量的概要，它展示了一个二阶张量对象，在三维的笛卡儿坐标系当中，张量的分量将形成矩阵。

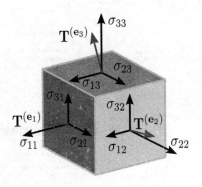

 参考图片来源于 tensor wiki：https://en.wikipedia.org/wiki/Tensor。

1.1.3 线性代数运算

以下将阐述线性代数的各种运算。

1. 向量

Norm 函数用于得到向量的大小，向量 x 的范数（norm）是对原点到点 x 的距离的度量。它也称为 L^p 范数，当 p=2 时，就是我们所熟知的**欧几里得范数**（Euclidean norm）。以下示例展示了如何计算给定向量的 L^p 范数：

```
import tensorflow as tf

vector = tf.constant([[4,5,6]], dtype=tf.float32)
eucNorm = tf.norm(vector, ord="euclidean")

with tf.Session() as sess:
    print(sess.run(eucNorm))
```

输出结果为 8.77496。

2. 矩阵

矩阵是一个二维数组,其中每个元素由两个索引(i, j)来标识,而不是一个。如果一个实矩阵 X 高为 m,宽为 n,则称 $X \in Rm \times n$,R 在这里是一组实数。

以下示例展示了如何将不同的矩阵转换为张量对象:

```
# convert matrices to tensor objects
import numpy as np
import tensorflow as tf

# create a 2x2 matrix in various forms
matrix1 = [[1.0, 2.0], [3.0, 40]]
matrix2 = np.array([[1.0, 2.0], [3.0, 40]], dtype=np.float32)
matrix3 = tf.constant([[1.0, 2.0], [3.0, 40]])

print(type(matrix1))
print(type(matrix2))
print(type(matrix3))

tensorForM1 = tf.convert_to_tensor(matrix1, dtype=tf.float32)
tensorForM2 = tf.convert_to_tensor(matrix2, dtype=tf.float32)
tensorForM3 = tf.convert_to_tensor(matrix3, dtype=tf.float32)

print(type(tensorForM1))
print(type(tensorForM2))
print(type(tensorForM3))
```

输出结果如下:

```
<class 'list'>
<class 'numpy.ndarray'>
<class 'tensorflow.python.framework.ops.Tensor'>
<class 'tensorflow.python.framework.ops.Tensor'>
<class 'tensorflow.python.framework.ops.Tensor'>
<class 'tensorflow.python.framework.ops.Tensor'>
```

3. 矩阵乘法

矩阵 A 和矩阵 B 相乘会得到第三个矩阵 C:

$$C = AB$$

矩阵元素的逐点相乘称为**哈达玛积**(Hadamard product),记为 $A \times B$。

两个向量 x 和 y 维数相同，则它们的点积是（转为矩阵）x 与（转为矩阵）y 转置的矩阵乘积。矩阵积 $C=AB$ 就像计算矩阵 A 的行 i 与矩阵 B 的列 j 的点积 $C_{i,j}$。

$$C_{i,j}=\sum_k A_{i,k}B_{k,j}$$

以下示例展示了使用张量对象的哈达玛积与点积：

```
import tensorflow as tf

mat1 = tf.constant([[4, 5, 6],[3,2,1]])
mat2 = tf.constant([[7, 8, 9],[10, 11, 12]])

# hadamard product (element wise)
mult = tf.multiply(mat1, mat2)

# dot product (no. of rows = no. of columns)
dotprod = tf.matmul(mat1, tf.transpose(mat2))

with tf.Session() as sess:
    print(sess.run(mult))
    print(sess.run(dotprod))
```

输出结果如下：

```
[[28 40 54][30 22 12]]
[[122 167][ 46 64]]
```

4. 迹运算

矩阵 A 的迹运算 Tr(A) 是对矩阵主对角线（从左上方至右下方的对角线）上的各个元素求总和。以下示例展示了如何对张量对象做迹运算：

```
import tensorflow as tf

mat = tf.constant([
 [0, 1, 2],
 [3, 4, 5],
 [6, 7, 8]
], dtype=tf.float32)

# get trace ('sum of diagonal elements') of the matrix
mat = tf.trace(mat)

with tf.Session() as sess:
    print(sess.run(mat))
```

输出结果为 12.0。

5. 矩阵转置

矩阵转置是矩阵沿主对角线上的镜像。矩阵等于它自身的转置的任意矩阵称为对称矩阵。

$$A = \begin{bmatrix} A_{1,1} & A_{1,2} \\ A_{2,1} & A_{2,2} \\ A_{3,1} & A_{3,2} \end{bmatrix} \Rightarrow A^{\mathrm{T}}$$

以下示例展示了如何对张量对象做转置运算：

```
import tensorflow as tf
x = [[1,2,3],[4,5,6]]
x = tf.convert_to_tensor(x)
xtrans = tf.transpose(x)

y=([[[1,2,3],[6,5,4]],[[4,5,6],[3,6,3]]])
y = tf.convert_to_tensor(y)
ytrans = tf.transpose(y, perm=[0, 2, 1])

with tf.Session() as sess:
   print(sess.run(xtrans))
   print(sess.run(ytrans))
```

输出结果如下：

[[1 4] [2 5] [3 6]]

6. 对角矩阵

对角矩阵是一个主对角线之外的元素皆为 0 的矩阵，只有主对角线元素可以为非零。不是所有的对角矩阵都是方阵。

对矩阵做对角运算，我们能得到一个给定矩阵的对角线，并创建一个给定对角线的矩阵，在 TensorFlow 中我们使用 diag 函数做对角运算。以下示例展示了如何在张量对象上进行对角运算：

```python
import tensorflow as tf

mat = tf.constant([
 [0, 1, 2],
 [3, 4, 5],
 [6, 7, 8]
], dtype=tf.float32)

# get diagonal of the matrix
diag_mat = tf.diag_part(mat)

# create matrix with given diagonal
mat = tf.diag([1,2,3,4])

with tf.Session() as sess:
   print(sess.run(diag_mat))
   print(sess.run(mat))
```

输出结果如下：

```
[ 0.  4.  8.]
[[1 0 0 0][0 2 0 0] [0 0 3 0] [0 0 0 4]]
```

7. 单位矩阵

单位矩阵 I 不会改变任意向量，以向量 V 为例，当与 I 相乘时，V 不变。

以下示例展示了对于给定大小如何生成单位矩阵：

```python
import tensorflow as tf

identity = tf.eye(3, 3)

with tf.Session() as sess:
   print(sess.run(identity))
```

输出结果如下：

```
[[ 1.  0.  0.] [ 0.  1.  0.] [ 0.  0.  1.]]
```

8. 逆矩阵

矩阵 I 的逆矩阵记为 I^{-1}。参考下面的等式，我们使用反函数和另一个不同的值 b 来求逆矩阵，对于 x 有多种求解方法。注意这个性质：

$$A^{-1}A = 1$$
$$Ax = b$$
$$A^{-1}Ax = a^{-1}b$$
$$I_n x = A^{-1}b$$
$$x = A^{-1}b$$

以下示例展示了如何使用 matrix_inverse 运算来计算逆矩阵:

```
import tensorflow as tf
mat = tf.constant([[2, 3, 4], [5, 6, 7], [8, 9, 10]], dtype=tf.float32)
print(mat)
inv_mat = tf.matrix_inverse(tf.transpose(mat))
with tf.Session() as sess:
    print(sess.run(inv_mat))
```

1.1.4 求解线性方程

TensorFlow 可以使用 solve 运算求解一系列线性方程组。让我们先不使用库中的函数来解释如何求解线性方程组,之后再使用 solve 来解释。

线性方程表示如下:

$$ax + b = y y - ax = b$$
$$y - ax = b$$
$$y/b - a/b(x) = 1$$

我们的工作是根据观察到的点来找出上述方程中 a 和 b 的值。首先,创建矩阵元素,第一列表示 x 的值,第二列表示 y 的值。

考虑到 X 是输入矩阵,A 是需要学习的参数,我们建立表达式 $AX = B$,因此,$A = BX^{-1}$。

下面的示例通过代码演示了如何求解线性方程:

$$3x+2y=15$$
$$4x-y=10$$

```python
import tensorflow as tf

# equation 1
x1 = tf.constant(3, dtype=tf.float32)
y1 = tf.constant(2, dtype=tf.float32)
point1 = tf.stack([x1, y1])

# equation 2
x2 = tf.constant(4, dtype=tf.float32)
y2 = tf.constant(-1, dtype=tf.float32)
point2 = tf.stack([x2, y2])

# solve for AX=C
X = tf.transpose(tf.stack([point1, point2]))
C = tf.ones((1,2), dtype=tf.float32)

A = tf.matmul(C, tf.matrix_inverse(X))

with tf.Session() as sess:
    X = sess.run(X)
    print(X)

    A = sess.run(A)
    print(A)

b = 1 / A[0][1]
a = -b * A[0][0]
print("Hence Linear Equation is: y = {a}x + {b}".format(a=a, b=b))
```

输出结果如下:

```
[[ 3.  4.][ 2. -1.]]
 [[ 0.27272728  0.09090909]]
Hence Linear Equation is: y = -2.9999999999999996x + 10.999999672174463
```

圆的正则方程是 $x2+y2+dx+ey+f=0$,为求解参数 d、e、f,我们使用 TensorFlow slove 运算来实现,代码如下:

```python
# canonical circle equation
# x2+y2+dx+ey+f = 0
# dx+ey+f=-(x2+y2) ==> AX = B
# we have to solve for d, e, f
```

```
points = tf.constant([[2,1], [0,5], [-1,2]], dtype=tf.float64)
X = tf.constant([[2,1,1], [0,5,1], [-1,2,1]], dtype=tf.float64)
B = -tf.constant([[5], [25], [5]], dtype=tf.float64)

A = tf.matrix_solve(X,B)

with tf.Session() as sess:
    result = sess.run(A)
    D, E, F = result.flatten()
    print("Hence Circle Equation is: x**2 + y**2 + {D}x + {E}y + {F} = 0".format(**locals()))
```

输出结果显示在下面的代码中:

```
Hence Circle Equation is: x**2 + y**2 + -2.0x + -6.0y + 5.0 = 0
```

1.1.5 奇异值分解

当把一个整数分解成它的主要因子时,我们能理解关于它的有用的性质。同样,当分解一个矩阵时,我们也能理解很多函数不直接显现的性质。有两种类型的矩阵分解,即特征值分解和奇异值分解。

每个实矩阵都有一个奇异值分解,但这并不适用于特征值分解。举例来说,如果一个矩阵不是方阵,就不能做特征分解,必须使用奇异值分解来替代。

奇异值分解(SVD)的数学表达式为三个矩阵 U、S 和 V 的乘积,其中,U 的维度为 $m*r$,S 的维度为 $r*r$,V 的维度为 $r*n$。

$$A_{m \times n} = USV^{-1}$$

以下示例展示了在文本数据上使用 TensorFlow svd 运算的代码:

```
import numpy as np
import tensorflow as tf
import matplotlib.pyplot as plts

path = "/neuralnetwork-programming/ch01/plots"

text = ["I", "like", "enjoy",
        "deep", "learning", "NLP", "flying", "."]
```

```
xMatrix = np.array([[0,2,1,0,0,0,0,0],
            [2,0,0,1,0,1,0,0],
            [1,0,0,0,0,0,1,0],
            [0,1,0,0,1,0,0,0],
            [0,0,0,1,0,0,0,1],
            [0,1,0,0,0,0,0,1],
            [0,0,1,0,0,0,0,1],
            [0,0,0,0,1,1,1,0]], dtype=np.float32)

X_tensor = tf.convert_to_tensor(xMatrix, dtype=tf.float32)

# tensorflow svd
with tf.Session() as sess:
    s, U, Vh = sess.run(tf.svd(X_tensor, full_matrices=False))

for i in range(len(text)):
    plts.text(U[i,0], U[i,1], text[i])

plts.ylim(-0.8,0.8)
plts.xlim(-0.8,2.0)
plts.savefig(path + '/svd_tf.png')

# numpy svd
la = np.linalg
U, s, Vh = la.svd(xMatrix, full_matrices=False)

print(U)
print(s)
print(Vh)

# write matrices to file (understand concepts)
file = open(path + "/matx.txt", 'w')
file.write(str(U))
file.write("\n")
file.write("=============")
file.write("\n")
file.write(str(s))
file.close()

for i in range(len(text)):
    plts.text(U[i,0], U[i,1], text[i])

plts.ylim(-0.8,0.8)
plts.xlim(-0.8,2.0)
plts.savefig(path + '/svd_np.png')
```

输出显示如下:

[[-5.24124920e-01 -5.72859168e-01   9.54463035e-02   3.83228481e-01
   -1.76963374e-01 -1.76092178e-01 -4.19185609e-01 -5.57702743e-02]
 [-5.94438076e-01   6.30120635e-01 -1.70207784e-01   3.10038358e-0

```
    1.84062332e-01  -2.34777853e-01   1.29535481e-01   1.36813134e-01]
 [ -2.56274015e-01   2.74017543e-01   1.59810841e-01   3.73903001e-16
   -5.78984618e-01   6.36550903e-01  -3.32297325e-16  -3.05414885e-01]
 [ -2.85637408e-01  -2.47912124e-01   3.54610324e-01  -7.31901303e-02
    4.45784479e-01   8.36141407e-02   5.48721075e-01  -4.68012422e-01]
 [ -1.93139315e-01   3.38495038e-02  -5.00790417e-01  -4.28462476e-01
    3.47110212e-01   1.55483231e-01  -4.68663752e-01  -4.03576553e-01]
 [ -3.05134684e-01  -2.93989003e-01  -2.23433599e-01  -1.91614240e-01
    1.27460942e-01   4.91219401e-01   2.09592804e-01   6.57535374e-01]
 [ -1.82489842e-01  -1.61027774e-01  -3.97842437e-01  -3.83228481e-01
   -5.12923241e-01  -4.27574426e-01   4.19185609e-01  -1.18313827e-01]
 [ -2.46898428e-01   1.57254755e-01   5.92991650e-01  -6.20076716e-01
   -3.21868137e-02  -2.31065080e-01  -2.59070963e-01   2.37976909e-01]]
[ 2.75726271   2.67824793   1.89221275   1.61803401   1.19154561   0.94833982
  0.61803401   0.56999218]
[[ -5.24124920e-01  -5.94438076e-01  -2.56274015e-01  -2.85637408e-01
   -1.93139315e-01  -3.05134684e-01  -1.82489842e-01  -2.46898428e-01]
 [  5.72859168e-01  -6.30120635e-01  -2.74017543e-01   2.47912124e-01
   -3.38495038e-02   2.93989003e-01   1.61027774e-01  -1.57254755e-01]
 [ -9.54463035e-02   1.70207784e-01  -1.59810841e-01  -3.54610324e-01
    5.00790417e-01   2.23433599e-01   3.97842437e-01  -5.92991650e-01]
 [  3.83228481e-01   3.10038358e-01  -2.22044605e-16  -7.31901303e-02
   -4.28462476e-01  -1.91614240e-01  -3.83228481e-01  -6.20076716e-01]
 [ -1.76963374e-01   1.84062332e-01  -5.78984618e-01   4.45784479e-01
    3.47110212e-01   1.27460942e-01  -5.12923241e-01  -3.21868137e-02]
 [  1.76092178e-01   2.34777853e-01  -6.36550903e-01  -8.36141407e-02
   -1.55483231e-01  -4.91219401e-01   4.27574426e-01   2.31065080e-01]
 [  4.19185609e-01  -1.29535481e-01  -3.33066907e-16  -5.48721075e-01
    4.68663752e-01  -2.09592804e-01  -4.19185609e-01   2.59070963e-01]
 [ -5.57702743e-02   1.36813134e-01  -3.05414885e-01  -4.68012422e-01
   -4.03576553e-01   6.57535374e-01  -1.18313827e-01   2.37976909e-01]]
```

以下是上述数据集 SVD 的图：

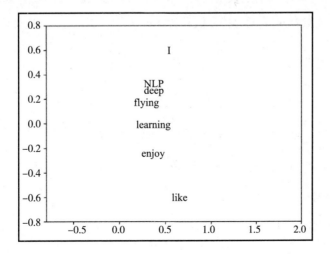

1.1.6 特征值分解

特征分解是比较常见的分解技术之一,通过特征分解,我们把矩阵分解成一组特征向量和特征值。

对于一个方阵而言,特征向量是一个向量 v,该向量与 A 相乘后,相当于对该向量进行缩放的非零向量 v。

$$Av = \lambda v$$

其中标量 λ 被称为与这个特征向量相对应的特征值。

A 的特征分解如下:

$$A = V \text{diag}(\lambda) V^{-1}$$

矩阵的特征分解描述了很多关于矩阵的有用信息。例如,当且仅当任何特征值都为 0 时,矩阵才是奇异的。

1.1.7 主成分分析

主成分分析(PCA) 将给定数据集投影到相对低维的线性空间上,以使投影数据方差最大化。PCA 需要数据矩阵 X 的特征值和特征向量。

数据矩阵 X 的 SVD 如下:

$$X = U \Sigma V^T$$
$$XX^T = (U \Sigma V^T)(U \Sigma V^T)^T$$
$$XX^T = (U \Sigma V^T)(V \Sigma U^T)$$
$$XX^T = U \Sigma^2 U^T$$

以下示例展示了使用 SVD 的 PCA:

```
import numpy as np
import tensorflow as tf
import matplotlib.pyplot as plt
import plotly.plotly as py
import plotly.graph_objs as go
import plotly.figure_factory as FF
import pandas as pd

path = "/neuralnetwork-programming/ch01/plots"
logs = "/neuralnetwork-programming/ch01/logs"

xMatrix = np.array([[0,2,1,0,0,0,0,0],
            [2,0,0,1,0,1,0,0],
            [1,0,0,0,0,0,1,0],
            [0,1,0,0,1,0,0,0],
            [0,0,0,1,0,0,0,1],
            [0,1,0,0,0,0,0,1],
            [0,0,1,0,0,0,0,1],
            [0,0,0,0,1,1,1,0]], dtype=np.float32)

def pca(mat):
    mat = tf.constant(mat, dtype=tf.float32)
    mean = tf.reduce_mean(mat, 0)
    less = mat - mean
    s, u, v = tf.svd(less, full_matrices=True, compute_uv=True)

    s2 = s ** 2
    variance_ratio = s2 / tf.reduce_sum(s2)

    with tf.Session() as session:
        run = session.run([variance_ratio])
    return run


if __name__ == '__main__':
    print(pca(xMatrix))
```

输出结果如下：

```
[array([  4.15949494e-01,   2.08390564e-01,   1.90929279e-01,
         8.36438537e-02,   5.55494241e-02,   2.46047471e-02,
         2.09326427e-02,   3.57540098e-16], dtype=float32)]
```

1.2 微积分

前几节涵盖了标准线性代数的一部分内容，但并未涵盖基础微积分。尽管我们使用的微积分相对简单，但其数学公式看起来可能非常复杂。在本中，我们用几个例子来介绍矩阵微积分的一些基本形式。

1.2.1 梯度

对于实数矩阵 A,其梯度函数定义为矩阵 A 的偏导数,形式如下:

$$\nabla_A f(A) \in \mathbb{R}^{m \times n} = \begin{bmatrix} \frac{\partial f(A)}{\partial A_{11}} & \frac{\partial f(A)}{\partial A_{12}} & \cdots & \frac{\partial f(A)}{\partial A_{1n}} \\ \frac{\partial f(A)}{\partial A_{21}} & \frac{\partial f(A)}{\partial A_{22}} & \cdots & \frac{\partial f(A)}{\partial A_{2n}} \\ \vdots & \vdots & \ddots & \vdots \\ \frac{\partial f(A)}{\partial A_{m1}} & \frac{\partial f(A)}{\partial A_{m2}} & \cdots & \frac{\partial f(A)}{\partial A_{mn}} \end{bmatrix}$$

$$(\nabla_A f(A))_{ij} = \frac{\partial f(A)}{\partial A_{ij}}$$

TensorFlow 不做数值微分,相反,它支持自动微分。通过在 Tensorflow 图中指定运算,它能自动执行图中的链式法则,因为它知道我们指定的每个运算的导数,将它们自动组合来进行运算。

以下示例使用 MNIST 数据集(由手写数字组成)来训练一个网络。该数据集的训练数据集有 60 000 个样本,测试数据集有 10 000 个样本。数字的图像尺寸做了归一化处理。

在这里,反向传播是在没有使用任何 API 的情况下执行的,导数手动计算。在 1000 个 测试样本中,我们得到了 913 个正确答案。这个概念将在下一章介绍。

以下代码片段描述了如何获取 mnist 数据集并初始化权重和偏置:

```
import tensorflow as tf

# get mnist dataset
from tensorflow.examples.tutorials.mnist import input_data
data = input_data.read_data_sets("MNIST_data/", one_hot=True)

# x represents image with 784 values as columns (28*28), y represents
```

```
output digit
x = tf.placeholder(tf.float32, [None, 784])
y = tf.placeholder(tf.float32, [None, 10])

# initialize weights and biases [w1,b1][w2,b2]
numNeuronsInDeepLayer = 30
w1 = tf.Variable(tf.truncated_normal([784, numNeuronsInDeepLayer]))
b1 = tf.Variable(tf.truncated_normal([1, numNeuronsInDeepLayer]))
w2 = tf.Variable(tf.truncated_normal([numNeuronsInDeepLayer, 10]))
b2 = tf.Variable(tf.truncated_normal([1, 10]))
```

现在定义一个两层的网络，每一层都有一个非线性的 sigmoid 函数。使用平方损失函数并用反向传播算法对其进行优化，如以下代码片段所示：

```
# non-linear sigmoid function at each neuron
def sigmoid(x):
    sigma = tf.div(tf.constant(1.0), tf.add(tf.constant(1.0), tf.exp(tf.negative(x))))
    return sigma

# starting from first layer with wx+b, then apply sigmoid to add non-linearity
z1 = tf.add(tf.matmul(x, w1), b1)
a1 = sigmoid(z1)
z2 = tf.add(tf.matmul(a1, w2), b2)
a2 = sigmoid(z2)

# calculate the loss (delta)
loss = tf.subtract(a2, y)

# derivative of the sigmoid function der(sigmoid)=sigmoid*(1-sigmoid)
def sigmaprime(x):
    return tf.multiply(sigmoid(x), tf.subtract(tf.constant(1.0), sigmoid(x)))

# backward propagation
dz2 = tf.multiply(loss, sigmaprime(z2))
db2 = dz2
dw2 = tf.matmul(tf.transpose(a1), dz2)

da1 = tf.matmul(dz2, tf.transpose(w2))
dz1 = tf.multiply(da1, sigmaprime(z1))
db1 = dz1
dw1 = tf.matmul(tf.transpose(x), dz1)

# finally update the network
eta = tf.constant(0.5)
step = [
    tf.assign(w1,
              tf.subtract(w1, tf.multiply(eta, dw1)))
```

```
    , tf.assign(b1,
           tf.subtract(b1, tf.multiply(eta,
                                   tf.reduce_mean(db1,
axis=[0]))))
    , tf.assign(w2,
           tf.subtract(w2, tf.multiply(eta, dw2)))
    , tf.assign(b2,
           tf.subtract(b2, tf.multiply(eta,
                                   tf.reduce_mean(db2,
axis=[0]))))
]

acct_mat = tf.equal(tf.argmax(a2, 1), tf.argmax(y, 1))
acct_res = tf.reduce_sum(tf.cast(acct_mat, tf.float32))

sess = tf.InteractiveSession()
sess.run(tf.global_variables_initializer())

for i in range(10000):
    batch_xs, batch_ys = data.train.next_batch(10)
    sess.run(step, feed_dict={x: batch_xs,
                              y: batch_ys})
    if i % 1000 == 0:
        res = sess.run(acct_res, feed_dict=
        {x: data.test.images[:1000],
         y: data.test.labels[:1000]})
        print(res)
```

输出结果如下：

```
Extracting MNIST_data
125.0
814.0
870.0
874.0
889.0
897.0
906.0
903.0
922.0
913.0
```

现在，我们使用 TensorFlow 来自动计算微分。以下示例使用 GradientDescent-Optimizer 来演示，在 1000 个测试样本中，我们得到了 924 个正确答案。

```
import tensorflow as tf

# get mnist dataset
from tensorflow.examples.tutorials.mnist import input_data
data = input_data.read_data_sets("MNIST_data/", one_hot=True)
```

```python
# x represents image with 784 values as columns (28*28), y represents
output digit
x = tf.placeholder(tf.float32, [None, 784])
y = tf.placeholder(tf.float32, [None, 10])

# initialize weights and biases [w1,b1][w2,b2]
numNeuronsInDeepLayer = 30
w1 = tf.Variable(tf.truncated_normal([784, numNeuronsInDeepLayer]))
b1 = tf.Variable(tf.truncated_normal([1, numNeuronsInDeepLayer]))
w2 = tf.Variable(tf.truncated_normal([numNeuronsInDeepLayer, 10]))
b2 = tf.Variable(tf.truncated_normal([1, 10]))

# non-linear sigmoid function at each neuron
def sigmoid(x):
    sigma = tf.div(tf.constant(1.0), tf.add(tf.constant(1.0), tf.exp(tf.negative(x))))
    return sigma

# starting from first layer with wx+b, then apply sigmoid to add non-linearity
z1 = tf.add(tf.matmul(x, w1), b1)
a1 = sigmoid(z1)
z2 = tf.add(tf.matmul(a1, w2), b2)
a2 = sigmoid(z2)

# calculate the loss (delta)
loss = tf.subtract(a2, y)

# derivative of the sigmoid function der(sigmoid)=sigmoid*(1-sigmoid)
def sigmaprime(x):
    return tf.multiply(sigmoid(x), tf.subtract(tf.constant(1.0), sigmoid(x)))

# automatic differentiation
cost = tf.multiply(loss, loss)
step = tf.train.GradientDescentOptimizer(0.1).minimize(cost)

acct_mat = tf.equal(tf.argmax(a2, 1), tf.argmax(y, 1))
acct_res = tf.reduce_sum(tf.cast(acct_mat, tf.float32))

sess = tf.InteractiveSession()
sess.run(tf.global_variables_initializer())

for i in range(10000):
    batch_xs, batch_ys = data.train.next_batch(10)
    sess.run(step, feed_dict={x: batch_xs,
                              y: batch_ys})
    if i % 1000 == 0:
        res = sess.run(acct_res, feed_dict=
        {x: data.test.images[:1000],
         y: data.test.labels[:1000]})
        print(res)
```

输出结果如下:

```
96.0
 777.0
 862.0
 870.0
 889.0
 901.0
 911.0
 905.0
 914.0
 924.0
```

以下示例展示了使用梯度下降的线性回归:

```
import tensorflow as tf
import numpy
import matplotlib.pyplot as plt
rndm = numpy.random

# config parameters
learningRate = 0.01
trainingEpochs = 1000
displayStep = 50

# create the training data
trainX = numpy.asarray([3.3,4.4,5.5,6.71,6.93,4.168,9.779,6.182,7.59,2.167,
                        7.042,10.791,5.313,7.997,5.654,9.27,3.12])
trainY =
numpy.asarray([1.7,2.76,2.09,3.19,1.694,1.573,3.366,2.596,2.53,1.221,
                        2.827,3.465,1.65,2.904,2.42,2.94,1.34])
nSamples = trainX.shape[0]

# tf inputs
X = tf.placeholder("float")
Y = tf.placeholder("float")

# initialize weights and bias
W = tf.Variable(rndm.randn(), name="weight")
b = tf.Variable(rndm.randn(), name="bias")

# linear model
linearModel = tf.add(tf.multiply(X, W), b)

# mean squared error
loss = tf.reduce_sum(tf.pow(linearModel-Y, 2))/(2*nSamples)

# Gradient descent
opt = tf.train.GradientDescentOptimizer(learningRate).minimize(loss)
```

```python
# initializing variables
init = tf.global_variables_initializer()

# run
with tf.Session() as sess:
    sess.run(init)

    # fitting the training data
    for epoch in range(trainingEpochs):
        for (x, y) in zip(trainX, trainY):
            sess.run(opt, feed_dict={X: x, Y: y})

        # print logs
        if (epoch+1) % displayStep == 0:
            c = sess.run(loss, feed_dict={X: trainX, Y:trainY})
            print("Epoch is:", '%04d' % (epoch+1), "loss=",
"{:.9f}".format(c), "W=", sess.run(W), "b=", sess.run(b))

    print("optimization done...")
    trainingLoss = sess.run(loss, feed_dict={X: trainX, Y: trainY})
    print("Training loss=", trainingLoss, "W=", sess.run(W), "b=",
sess.run(b), '\n')

    # display the plot
    plt.plot(trainX, trainY, 'ro', label='Original data')
    plt.plot(trainX, sess.run(W) * trainX + sess.run(b), label='Fitted
line')
    plt.legend()
    plt.show()

    # Testing example, as requested (Issue #2)
    testX = numpy.asarray([6.83, 4.668, 8.9, 7.91, 5.7, 8.7, 3.1, 2.1])
    testY = numpy.asarray([1.84, 2.273, 3.2, 2.831, 2.92, 3.24, 1.35,
1.03])

    print("Testing... (Mean square loss Comparison)")
    testing_cost = sess.run(
        tf.reduce_sum(tf.pow(linearModel - Y, 2)) / (2 * testX.shape[0]),
        feed_dict={X: testX, Y: testY})
    print("Testing cost=", testing_cost)
    print("Absolute mean square loss difference:", abs(trainingLoss -
testing_cost))

    plt.plot(testX, testY, 'bo', label='Testing data')
    plt.plot(trainX, sess.run(W) * trainX + sess.run(b), label='Fitted
line')
    plt.legend()
    plt.show()
```

输出结果如下：

```
Epoch is: 0050 loss= 0.141912043 W= 0.10565 b= 1.8382
Epoch is: 0100 loss= 0.134377643 W= 0.11413 b= 1.7772
Epoch is: 0150 loss= 0.127711013 W= 0.122106 b= 1.71982
Epoch is: 0200 loss= 0.121811897 W= 0.129609 b= 1.66585
Epoch is: 0250 loss= 0.116592340 W= 0.136666 b= 1.61508
Epoch is: 0300 loss= 0.111973859 W= 0.143304 b= 1.56733
Epoch is: 0350 loss= 0.107887231 W= 0.149547 b= 1.52241
Epoch is: 0400 loss= 0.104270980 W= 0.15542 b= 1.48017
Epoch is: 0450 loss= 0.101070963 W= 0.160945 b= 1.44043
Epoch is: 0500 loss= 0.098239250 W= 0.166141 b= 1.40305
Epoch is: 0550 loss= 0.095733419 W= 0.171029 b= 1.36789
Epoch is: 0600 loss= 0.093516059 W= 0.175626 b= 1.33481
Epoch is: 0650 loss= 0.091553882 W= 0.179951 b= 1.3037
Epoch is: 0700 loss= 0.089817807 W= 0.184018 b= 1.27445
Epoch is: 0750 loss= 0.088281371 W= 0.187843 b= 1.24692
Epoch is: 0800 loss= 0.086921677 W= 0.191442 b= 1.22104
Epoch is: 0850 loss= 0.085718453 W= 0.194827 b= 1.19669
Epoch is: 0900 loss= 0.084653646 W= 0.198011 b= 1.17378
Epoch is: 0950 loss= 0.083711281 W= 0.201005 b= 1.15224
Epoch is: 1000 loss= 0.082877308 W= 0.203822 b= 1.13198
optimization done...
Training loss= 0.0828773 W= 0.203822 b= 1.13198
Testing... (Mean square loss Comparison)
Testing cost= 0.0957726
Absolute mean square loss difference: 0.0128952
```

图如下所示：

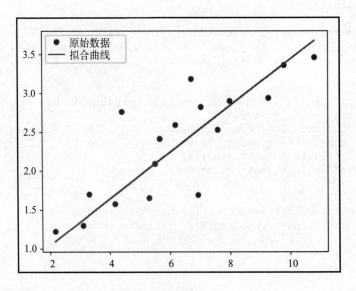

下图展示了使用模型在测试集上的拟合曲线：

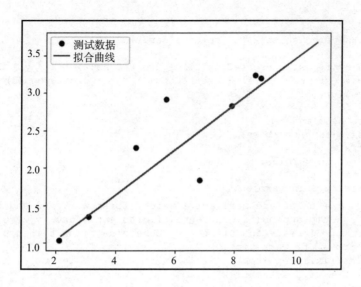

1.2.2 Hessian 矩阵

梯度是向量函数的一阶导数，而 Hessian 是二阶导数，我们现在将其标记为：

$$\nabla_x^2 f(x) \in \mathbb{R}^{n \times m} = \begin{bmatrix} \dfrac{\partial^2 f(x)}{\partial x_1^2} & \dfrac{\partial^2 f(x)}{\partial x_1 \partial x_2} & \cdots & \dfrac{\partial^2 f(x)}{\partial x_1 \partial x_n} \\ \dfrac{\partial^2 f(x)}{\partial x_2 \partial x_1} & \dfrac{\partial^2 f(x)}{\partial x_2^2} & \cdots & \dfrac{\partial^2 f(x)}{\partial x_2 \partial x_n} \\ \vdots & \vdots & \ddots & \vdots \\ \dfrac{\partial^2 f(x)}{\partial x_n \partial x_1} & \dfrac{\partial^2 f(x)}{\partial x_n \partial x_2} & \cdots & \dfrac{\partial^2 f(x)}{\partial x_n^2} \end{bmatrix}$$

与梯度类似，仅当 $f(x)$ 为实矩阵时才定义 Hessian。

 使用的代数函数为 $q(x) = x_1^2 + 2x_1 x_2 + 3x_2^2 + 4x_1 + 5x_2 + 6$。

以下示例展示了基于 TensorFlow 的 Hessian 实现：

```
import tensorflow as tf
import numpy as np

X = tf.Variable(np.random.random_sample(), dtype=tf.float32)
y = tf.Variable(np.random.random_sample(), dtype=tf.float32)
```

```
def createCons(x):
    return tf.constant(x, dtype=tf.float32)

function = tf.pow(X, createCons(2)) + createCons(2) * X * y + createCons(3)
* tf.pow(y, createCons(2)) + createCons(4) * X + createCons(5) * y +
createCons(6)

# compute hessian
def hessian(func, varbles):
    matrix = []
    for v_1 in varbles:
        tmp = []
        for v_2 in varbles:
            # calculate derivative twice, first w.r.t v2 and then w.r.t v1
            tmp.append(tf.gradients(tf.gradients(func, v_2)[0], v_1)[0])
        tmp = [createCons(0) if t == None else t for t in tmp]
        tmp = tf.stack(tmp)
        matrix.append(tmp)
    matrix = tf.stack(matrix)
    return matrix

hessian = hessian(function, [X, y])

sess = tf.Session()
sess.run(tf.initialize_all_variables())
print(sess.run(hessian))
```

输出结果如下：

```
[[ 2.  2.] [ 2.  6.]]
```

1.2.3 行列式

行列式向我们展示了有关矩阵的信息，这些信息有助于线性方程式，也有助于找到矩阵的逆。

对于一个给定的矩阵 X，行列式如下：

$$X = \begin{matrix} a & b & c \\ d & e & f \\ g & h & i \end{matrix}$$

$$\det(X) = a(ei-fh) - b(di-fg) - c(dh-eg)$$

以下示例展示了如何使用 TensorFlow 得到行列式：

```
import tensorflow as tf
import numpy as np

x = np.array([[10.0, 15.0, 20.0], [0.0, 1.0, 5.0], [3.0, 5.0, 7.0]],
dtype=np.float32)

det = tf.matrix_determinant(x)

with tf.Session() as sess:
    print(sess.run(det))
```

输出结果如下：

-15.0

1.3 最优化

作为深度学习的一部分，大多数情况下，我们希望优化一个函数的值，对于 x，最大化或者最小化 $f(x)$。诸如最小二乘、逻辑回归和支持向量机都是最优化问题的实例。其中许多技术将在后面的章节中详细介绍。

优化器

在这里，我们将学习 `AdamOptimizer`，TensorFlow `AdamOptimizer` 使用 Kingma 和 Ba 的 Adam 算法来控制学习率。相比简单的 `GradientDescentOptimizer`，`AdamOptimizer` 有很多优点。首先，它使用参数的移动平均值，这使其能够使用更大的步长，并在无须任何微调（fine-tuning）的情况下使步长收敛。

Adam 的缺点是在每一步训练中需要对每个参数进行更多的计算。也可以使用 `GradientDescentOptimizer`，但是在其快速收敛之前，它需要做更多的超参数调整。

以下示例展示了如何使用 Adam Optimizer：

- `tf.train.Optimizer` 创建一个优化器。
- `tf.train.Optimizer.minimize(loss,var_list)` 添加优化器运算到计算图中。

在 Tensorflow 中，不需要用户输入，自动微分会自动计算梯度。

```python
import numpy as np
import seaborn
import matplotlib.pyplot as plt
import tensorflow as tf

# input dataset
xData = np.arange(100, step=.1)
yData = xData + 20 * np.sin(xData/10)

# scatter plot for input data
plt.scatter(xData, yData)
plt.show()
# defining data size and batch size
nSamples = 1000
batchSize = 100

# resize
xData = np.reshape(xData, (nSamples,1))
yData = np.reshape(yData, (nSamples,1))

# input placeholders
x = tf.placeholder(tf.float32, shape=(batchSize, 1))
y = tf.placeholder(tf.float32, shape=(batchSize, 1))

# init weight and bias
with tf.variable_scope("linearRegression"):
 W = tf.get_variable("weights", (1, 1),
initializer=tf.random_normal_initializer())
 b = tf.get_variable("bias", (1,),
initializer=tf.constant_initializer(0.0))

 y_pred = tf.matmul(x, W) + b
 loss = tf.reduce_sum((y - y_pred)**2/nSamples)

# optimizer
opt = tf.train.AdamOptimizer().minimize(loss)
with tf.Session() as sess:
    sess.run(tf.global_variables_initializer())

    # gradient descent loop for 500 steps
    for _ in range(500):
     # random minibatch
     indices = np.random.choice(nSamples, batchSize)
```

```
        X_batch, y_batch = xData[indices], yData[indices]

        # gradient descent step
        _, loss_val = sess.run([opt, loss], feed_dict={x: X_batch, y:
y_batch})
```

下图为数据集的散点图:

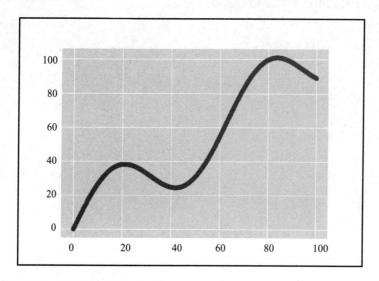

下图为数据集上学习的模型图:

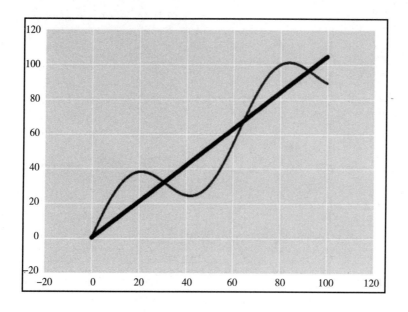

1.4 总结

在本章中，我们已经介绍了一些数学概念（这些是理解神经网络的关键）并回顾了与张量相关的数学知识。我们还演示了如何在 TensorFlow 中执行数学运算。在接下来的章节中，我们将反复利用这些概念进行实践应用。

CHAPTER 2
第 2 章

深度前馈神经网络

在第 1 章，我们了解了神经网络背后隐藏的数学原理。本章将重点关注最基本的神经网络模型，即**前馈神经网络**。我们也会介绍深度前馈网络。这种网络采用多个隐藏层来提升模型的准确率。

在本章中，我们将涵盖如下内容：

- 定义前馈神经网络。
- 理解反向传播。
- 在 TensorFlow 中实现前馈网络。
- 分析 Iris 数据集。
- 构建前馈网络对图片进行分类。

2.1 定义前馈神经网络

深度前馈网络，即前馈神经网络，有时也称为**多层感知机**（Multilayer Perceptron，MLP）。前馈神经网络的目标是逼近函数 $f^*(x)$。例如，对于一个分类器，我们记 $y=f^*(x)$ 为输入 x 到标签 y 的真实映射关系，$y=f(x;\theta)$ 为前馈网络定义的映射关系。其中 θ 即为模型需要学习的参数，学习的目标是使 $f(x;\theta)$ 为 $f^*(x)$ 的最佳近似。

我们将在第 5 章讨论递归神经网络（Recurrent Neural Network，RNN）。前馈神经网络是递归神经网络的基础,递归神经网络在自然语言的很多应用中有着重要的作用。前馈神经网络被称为网络的原因是它们将很多不同的表示它们的函数进行组合。我们将这些函数组合为有向无环图。

前馈神经网络模型和一个描述函数如何组合在一起的有向无环图有关。例如有3个函数 $f(1), f(2), f(3)$,将它们进行连接组成函数 $f(x)=f(3)(f(2)(f(1)(x)))$。这种链式的结构是神经网络中最常用的结构。在本例中,$f(1)$ 称为网络的**第一层**,$f(2)$ 称为**第二层**,以此类推。链式结构的总长度就是模型的深度。深度学习这个名字就是由此而来。前馈网络的最后一层称为**输出层**。

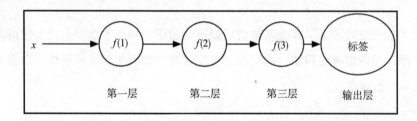

图 2-1 在输入 x 上激活各种函数以形成神经网络的图示

这些网络被称为"神经"网络,因为人们受神经科学的启发而提出了这种结构。每一个隐藏层都是一个向量。这些隐藏层的维度决定了模型的宽度。

2.2 理解反向传播

当使用前馈神经网络接受输入 x,并产生输出 $\hat{y}$ 时,信息通过网络中的元素向前传播。输入 x 提供信息然后传递给每一层的隐藏层,最后产生输出 $\hat{y}$。这一过程称为**前向传播**。在训练的时候,前向传播一直进行到网络产生了损失 $J(\theta)$ 为止,$J(\theta)$ 是标量。反向传播算法通常称为反向传播,允许信息从损失开始,通过网络向后传播,以此来计算梯度。

我们可以直接计算梯度的解析表达式,但是数值计算这样的表达式在计算上很可能非常昂贵。反向传播算法采用了这样一种简单而且(在计算上)不昂贵的方式来计算梯度。

反向传播仅仅指的是计算梯度的方法,而其他算法(例如随机梯度下降法)是指在实际中采用的技术。

2.3 在 TensorFlow 中实现前馈神经网络

在 TensorFlow 中,通过为隐藏层定义占位符,计算激活函数的值,然后使用它们计算预测值,我们可以很方便地实现前馈神经网络。我们看一个使用前馈神经网络进行分类的例子:

```
X = tf.placeholder("float", shape=[None, x_size])
y = tf.placeholder("float", shape=[None, y_size])
weights_1 = initialize_weights((x_size, hidden_size), stddev)
weights_2 = initialize_weights((hidden_size, y_size), stddev)
sigmoid = tf.nn.sigmoid(tf.matmul(X, weights_1))
y = tf.matmul(sigmoid, weights_2)
```

一旦定义了预测值的张量,就可以计算 `cost` 函数:

```
cost = tf.reduce_mean(tf.nn.OPERATION_NAME(labels=<actual value>,
logits=<predicted value>))
updates_sgd = tf.train.GradientDescentOptimizer(sgd_step).minimize(cost)
```

这里,`OPERATION_NAME` 可以是以下任意一个:

❑ `tf.nn.sigmoid_cross_entropy_with_logits`:计算 `logits` 和 `labels` 的交叉熵。

```
  sigmoid_cross_entropy_with_logits(
    _sentinel=None,
    labels=None,
    logits=None,
    name=None
  )Formula implemented is max(x, 0) - x * z + log(1 + exp(-abs(x)))
```

  `_sentinel`:用于避免使用位置参数。内部,不要使用。

  `labels`:和 `logits` 大小、类型一样的张量。

  `logits`:`float32` 或者 `float64` 的张量。计算公式为 ($x =$ `logits`, $z =$ `labels`) $\max(x,0) - x*z + \log(1 + \exp(-\text{abs}(x)))$。

- tf.nn.softmax：对输入的张量使用softmax激活函数。这只是归一化操作，确保张量中行的所有概率加起来为1。它不能直接用于分类。

  ```
  softmax = exp(logits) / reduce_sum(exp(logits), dim)
  ```

  logits：非空张量。必需是以下类型之一——half、float32或者float64。

  dim：进行softmax归一的维度。默认为-1，即在最后一个维度上进行softmax归一。

  name：这个操作的名称（可选）。

- tf.nn.log_softmax：计算softmax函数的对数值，这对归一化的不足有帮助。这也只是一个归一化函数。

  ```
  log_softmax(
    logits,
    dim=-1,
    name=None
  )
  ```

  logits：非空张量。必需是以下类型之一——half、float32或者float64。

  dim：进行softmax归一的维度。默认为-1，即在最后一个维度上进行softmax归一。

  name：这个操作的名称（可选）。

- tf.nn.softmax_cross_entropy_with_logits

  ```
  softmax_cross_entropy_with_logits(
    _sentinel=None,
    labels=None,
    logits=None,
    dim=-1,
    name=None
  )
  ```

  _sentinel：用于避免使用位置参数。内部，不要使用。

labels：每一行 labels[i] 必须是有效的概率分布。

logits：未归一化的对数概率。

dim：类别所在维度。默认为 –1，即最后一个维度。

name：这个操作的名称（可选）。

上面的代码段计算了 logits 和 labels 之间的 softmax 交叉熵。虽然类别之间是相互排斥的，但它们的概率却不一定如此。所以我们必须确保 labels 的每一行是一个有效的概率分布。

对于互斥的 labels，我们使用 sparse_softmax_cross_entropy_with_logits。（这时，一次只有一个类别是真的，即只有一个是 1，其他都是 0）。

❑ tf.nn.sparse_softmax_cross_entropy_with_logits

```
sparse_softmax_cross_entropy_with_logits(
  _sentinel=None,
  labels=None,
  logits=None,
  name=None
)
```

labels：大小为 [d_0, d_1, …, d_(r-1)]（r 是 lables 和结果的秩），类型为 int32 或者 int64 的张量。labels 中的每一个元素必须是 [0,num_classes) 的索引。其他值会引起错误，如果在 CPU 上运行，那么对应的损失会是 NAN，如果在 GPU 上运行，那么对应的梯度会变为 NAN。

logits：大小为 [d_0, d_1, …, d_(r-1), num_classes]，类型为 float32 或者 float64 的未归一化的对数概率。

上面的代码计算了 logits 和 labels 之间的稀疏 softmax 交叉熵。我们认为给定标签的概率是互斥的。软分类（概率非互斥）是不允许的，标签向量必须为 logits 的每一行提供一个真实类别的单一索引。

❏ tf.nn.weighted_cross_entropy_with_logits

```
weighted_cross_entropy_with_logits(
  targets,
  logits,
  pos_weight,
  name=None
)
```

targets：和 logits 大小、类型一样的张量。

logits：float32 或者 float64 的张量。

pos_weight：正样本使用的系数（权重）。

这个函数与 sigmoid_cross_entropy_with_logits() 类似，只是 pos_weight 使得我们可以提高或下调正样本错判为负样本的损失和负样本错判为正样本的损失之间的权重，从而在召回率和精度之间进行权衡。

2.4 分析 Iris 数据集

我们看一个使用 Iris 数据集的前馈网络的例子。

 你可以从 https://github.com/ml-resources/neuralnetwork-programming/blob/ed1/ch02/iris/iris.csv 下载数据集，从 https://github.com/ml-resources/neuralnetwork-programming/blob/ed1/ch02/iris/target.csv 下载对应标签。

在 Iris 数据集中，我们将使用 150 行数据。这些数据由三种鸢尾属植物（山鸢尾（Iris setosa）、维吉尼亚鸢尾（Iris virginica）和杂色鸢尾（Iris versicolor）组成，每个类别 50 个样本。

三种鸢尾属的花瓣的比较如下图所示。

| 山鸢尾 | 杂色鸢尾 | 维吉尼亚鸢尾 |

在数据集中,每行包含每朵花的以下数据:萼片长度、萼片宽度、花瓣长度、花瓣宽度和花种类。花种类以整数形式存储,其中 0 表示山鸢尾,1 表示杂色鸢尾,2 表示维吉尼亚鸢尾。

首先,创建一个 run() 函数,这个函数有 3 个参数——隐藏层大小 h_size、权重的标准差 stddev 和随机梯度下降的步长大小(学习率)sgd_step:

```
def run(h_size, stddev, sgd_step)
```

我们使用 numpy 中的 genfromtxt 函数来读取数据。读入的 Iris 数据的大小为 150 行 4 列。数据存储在 all_X 变量中。目标标签从 target.csv 中读取,存储在 all_Y 变量中,大小为 150 行 3 列:

```
def load_iris_data():
    from numpy import genfromtxt
    data = genfromtxt('iris.csv', delimiter=',')
    target = genfromtxt('target.csv', delimiter=',').astype(int)
    # Prepend the column of 1s for bias
    L, W  = data.shape
    all_X = np.ones((L, W + 1))
    all_X[:, 1:] = data
    num_labels = len(np.unique(target))
    all_y = np.eye(num_labels)[target]
    return train_test_split(all_X, all_y, test_size=0.33,
random_state=RANDOMSEED)
```

数据读取完毕后,我们根据 x_size、y_size 以及传递给 run() 函数的隐藏层大小 h_size 和标准差 stddev 初始化权重矩阵。

- x_size = 5
- y_size = 3
- h_size = 128（或为隐藏层中的神经元选择的其他数字）

```
# Size of Layers
x_size = train_x.shape[1] # Input nodes: 4 features and 1 bias
y_size = train_y.shape[1] # Outcomes (3 iris flowers)
# variables
X = tf.placeholder("float", shape=[None, x_size])
y = tf.placeholder("float", shape=[None, y_size])
weights_1 = initialize_weights((x_size, h_size), stddev)
weights_2 = initialize_weights((h_size, y_size), stddev)
```

接下来，我们将使用在 forward_propagation() 函数中定义的 sigmoid 函数作为激活数进行预测：

```
def forward_propagation(X, weights_1, weights_2):
    sigmoid = tf.nn.sigmoid(tf.matmul(X, weights_1))
    y = tf.matmul(sigmoid, weights_2)
    return y
```

首先，sigmoid 的输出是通过计算输入 X 和 weights_1 得到的。之后，我们将 sigmoid 的输出和 weights_2 进行矩阵相乘运算，得到 y。

```
y_pred = forward_propagation(X, weights_1, weights_2)
predict = tf.argmax(y_pred, dimension=1)
```

接着，我们使用梯度下降定义损失函数和优化方法。首先看一下使用的 GradientDescentOptimizer 方法。这个方法在 tf.train.GradientDescentOptimizer 类中定义，它采用梯度下降算法。

我们用如下构造函数来创建一个实例，并将 sgd_step 作为一个参数传递：

```
# constructor for GradientDescentOptimizer
__init__(
  learning_rate,
  use_locking=False,
  name='GradientDescent'
)
```

参数的含义如下：

- `learning_rate`：张量或者浮点数，表示使用的学习率。
- `use_locking`：如果设为 `ture`，更新操作会加锁。
- `name`：执行梯度下降时产生操作的名称前缀，可选。默认的名称是 `"Gradient Descent"`。

以下是实现 cost 函数的代码：

```
cost = tf.reduce_mean(tf.nn.softmax_cross_entropy_with_logits(labels=y, logits=y_pred))
updates_sgd = tf.train.GradientDescentOptimizer(sgd_step).minimize(cost)
```

接下来我们将执行以下步骤：

1）初始化 TensorFlow 会话：

```
sess = tf.Session()
```

2）使用 `tf.initialize_all_variables()` 初始化所有变量；返回的对象用于实例化会话。

3）进行迭代（1 到 50）。

4）对 `train_x` 和 `train_y` 中的每一次迭代，执行 `updates_sgd`。

5）计算 `train_accuracy` 和 `test_accuracy`。

我们把每一次迭代的准确率存储在列表中，这样就可以绘制一个图。

```
init = tf.initialize_all_variables()
steps = 50
sess.run(init)
x   = np.arange(steps)
test_acc = []
train_acc = []
print("Step, train accuracy, test accuracy")
for step in range(steps):
    # Train with each example
    for i in range(len(train_x)):
        sess.run(updates_sgd, feed_dict={X: train_x[i: i + 1], y: train_y[i: i + 1]})

    train_accuracy = np.mean(np.argmax(train_y, axis=1) ==
```

```
                              sess.run(predict, feed_dict={X: train_x,
y: train_y}))
        test_accuracy = np.mean(np.argmax(test_y, axis=1) ==
                              sess.run(predict, feed_dict={X: test_x, y:
test_y}))

        print("%d, %.2f%%, %.2f%%"
            % (step + 1, 100. * train_accuracy, 100. * test_accuracy))
        test_acc.append(100. * test_accuracy)
        train_acc.append(100. * train_accuracy)
```

代码执行

将 h_size 设为 128，标准差设为 0.1，sgd_step 设为 0.01，以执如下代码：

```
def run(h_size, stddev, sgd_step):
 ...

def main():
run(128,0.1,0.01)

if __name__ == '__main__':
main()
```

之前的代码会输出如下图片，该图展示了每一步中的训练集准确率和测试集准确率。

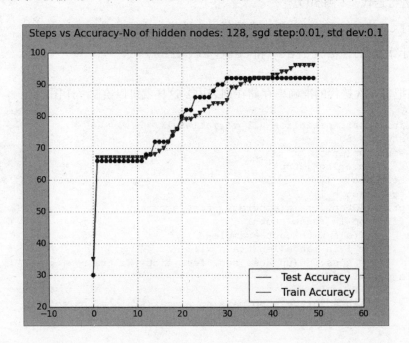

我们来比较一下 SGD 步长变化对训练准确率的影响。以下代码与之前的示例代码非常相似，但是我们将使用多个 SGD 步长重新运行代码，以此了解 SGD 步长对准确率的影响程度。

```
def run(h_size, stddev, sgd_steps):
    ....
    test_accs = []
    train_accs = []
    time_taken_summary = []
    for sgd_step in sgd_steps:
        start_time = time.time()
        updates_sgd = tf.train.GradientDescentOptimizer(sgd_step).minimize(cost)
        sess = tf.Session()
        init = tf.initialize_all_variables()
        steps = 50
        sess.run(init)
        x = np.arange(steps)
        test_acc = []
        train_acc = []

        print("Step, train accuracy, test accuracy")


        for step in range(steps):
            # Train with each example
            for i in range(len(train_x)):
                sess.run(updates_sgd, feed_dict={X: train_x[i: i + 1],
                    y: train_y[i: i + 1]})

            train_accuracy = np.mean(np.argmax(train_y, axis=1) ==
                                    sess.run(predict,
                                    feed_dict={X: train_x, y: train_y}))
            test_accuracy = np.mean(np.argmax(test_y, axis=1) ==
                                    sess.run(predict,
                                    feed_dict={X: test_x, y: test_y}))

            print("%d, %.2f%%, %.2f%%"
                  % (step + 1, 100. * train_accuracy, 100. * test_accuracy))
            #x.append(step)
            test_acc.append(100. * test_accuracy)
            train_acc.append(100. * train_accuracy)
        end_time = time.time()
        diff = end_time -start_time
        time_taken_summary.append((sgd_step,diff))
        t = [np.array(test_acc)]
        t.append(train_acc)
        train_accs.append(train_acc)
```

对每一个 SGD 步长值，之前代码都会分别输出一个训练集和测试集准确率的数组。在本例中，我们用 [0.01,0.02,0.03] 这组步长值调用 sgd_steps 函数。

```
def main():
    sgd_steps = [0.01,0.02,0.03]
    run(128,0.1,sgd_steps)

if __name__ == '__main__':
    main()
```

下图展示了训练准确率是如何随着 sgd_steps 变化的。对于值为 0.03 的 SGD 步长值，由于步长较大，它可以更快地达到一个更高的准确率。

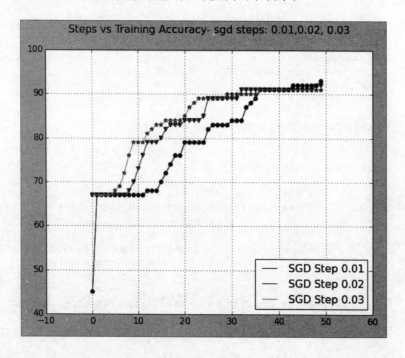

2.5 使用前馈网络进行图像分类

现在我们来看看如何使用前馈网络对图像进行分类。我们将使用 notMNIST 数据集。该数据集由九个字母的图像组成，从 A 到 I。notMNIST 数据集类似于 MNIST 数据集，但是它专注于字母而不是数字（http://yaroslavvb.blogspot.in/2011/09/notmnist-dataset.html）。

我们已经将原始数据集缩小为较小版本以便于进行训练。下载 ZIP 文件并将其提取到数据集所在的文件夹 (`https://1drv.ms/f/s!Av6fk5nQi2j-kniw-8GtP8sdWejs`)。

Python 的 pickle 模块实现了一个对 Python 对象结构进行序列化和反序列化的算法。pickling 是一个将 Python 对象层次结构转换为字节流的过程，unpickling 是反向操作，将字节流转换回对象层次结构。pickling（或者 unpickling）也称为 **serialization**、**marshaling** 或者 **flattening**。

首先，我们使用 `maybe_pickle()` 在 `numpy.ndarray` 中加载下面文件列表中的图像。

```
test_folders = ['./notMNIST_small/A', './notMNIST_small/B',
'./notMNIST_small/C', './notMNIST_small/D',
'./notMNIST_small/E', './notMNIST_small/F', './notMNIST_small/G',
'./notMNIST_small/H',
'./notMNIST_small/I', './notMNIST_small/J']
train_folders = ['./notMNIST_large_v2/A', './notMNIST_large_v2/B',
'./notMNIST_large_v2/C', './notMNIST_large_v2/D',
'./notMNIST_large_v2/E', './notMNIST_large_v2/F', './notMNIST_large_v2/G',
'./notMNIST_large_v2/H',
'./notMNIST_large_v2/I', './notMNIST_large_v2/J']
maybe_pickle(data_folders, min_num_images_per_class, force=False):
```

`maybe_pickle` 使用 `load_letter` 方法将单个文件夹中的图像读取为 `ndarray`。

```
def load_letter(folder, min_num_images):
  image_files = os.listdir(folder)
  dataset = np.ndarray(shape=(len(image_files), image_size, image_size),
                       dtype=np.float32)
  num_images = 0
  for image in image_files:
    image_file = os.path.join(folder, image)
    try:
      image_data = (ndimage.imread(image_file).astype(float) -
                    pixel_depth / 2) / pixel_depth
      if image_data.shape != (image_size, image_size):
        raise Exception('Unexpected image shape: %s' %
str(image_data.shape))
      dataset[num_images, :, :] = image_data
      num_images = num_images + 1
    except IOError as e:
      print('Could not read:', image_file, ':', e, '- it\'s ok, skipping.')
  dataset = dataset[0:num_images, :, :]
  if num_images < min_num_images:
    raise Exception('Fewer images than expected: %d < %d' %
                    (num_images, min_num_images))
  print('Dataset tensor:', dataset.shape)
```

```
print('Mean:', np.mean(dataset))
print('Standard deviation:', np.std(dataset))
return dataset
```

maybe_pickle方法会被调用2次，分别用于读取训练文件夹中的图像和测试文件夹中的图像。

```
train_datasets = maybe_pickle(train_folders, 100)
test_datasets = maybe_pickle(test_folders, 50)
```

输出类似于下面的截图。

第一张截图显示了dataset_names列表中变量的值：

```
▼ ≣ dataset_names = {list} <type 'list'>: ['./notMNIST_
    ⓘ __len__ = {int} 10
    ⓘ 00 = {str} './notMNIST_large_v2/A.pickle'
    ⓘ 01 = {str} './notMNIST_large_v2/B.pickle'
    ⓘ 02 = {str} './notMNIST_large_v2/C.pickle'
    ⓘ 03 = {str} './notMNIST_large_v2/D.pickle'
    ⓘ 04 = {str} './notMNIST_large_v2/E.pickle'
    ⓘ 05 = {str} './notMNIST_large_v2/F.pickle'
    ⓘ 06 = {str} './notMNIST_large_v2/G.pickle'
    ⓘ 07 = {str} './notMNIST_large_v2/H.pickle'
    ⓘ 08 = {str} './notMNIST_large_v2/I.pickle'
    ⓘ 09 = {str} './notMNIST_large_v2/J.pickle'
```

之后的截图显示notMNIST_small数据集的dataset_names变量的值：

```
▼ ≣ dataset_names = {list} <type 'list'>: ['./not
    ⓘ __len__ = {int} 10
    ⓘ 00 = {str} './notMNIST_small/A.pickle'
    ⓘ 01 = {str} './notMNIST_small/B.pickle'
    ⓘ 02 = {str} './notMNIST_small/C.pickle'
    ⓘ 03 = {str} './notMNIST_small/D.pickle'
    ⓘ 04 = {str} './notMNIST_small/E.pickle'
    ⓘ 05 = {str} './notMNIST_small/F.pickle'
    ⓘ 06 = {str} './notMNIST_small/G.pickle'
    ⓘ 07 = {str} './notMNIST_small/H.pickle'
    ⓘ 08 = {str} './notMNIST_small/I.pickle'
    ⓘ 09 = {str} './notMNIST_small/J.pickle'
```

接着，调用merge_datasets函数。这个函数将每个字母的pickle文件并入以下 ndarray：

- valid_dataset
- valid_labels
- train_dataset
- train_labels

```
train_size = 1000
valid_size = 500
test_size = 500

valid_dataset, valid_labels, train_dataset, train_labels = merge_datasets(
    train_datasets, train_size, valid_size)
_, _, test_dataset, test_labels = merge_datasets(test_datasets,
test_size)
```

之前代码的输出如下所示：

```
Training dataset and labels shape: (1000, 28, 28) (1000,)
Validation dataset and labels shape: (500, 28, 28) (500,)
Testing dataset and labels shape: (500, 28, 28) (500,)
```

最后，我们以键值对的形式将 ndarray 存储在 noMNIST.pickle 文件中，其中键是 train_dataset、train_labels、valid_dataset、valid_labels、test_dataset 和 test_labels，值是各自的 ndarray，如下面的代码所示：

```
try:
  f = open(pickle_file, 'wb')
  save = {
    'train_dataset': train_dataset,
    'train_labels': train_labels,
    'valid_dataset': valid_dataset,
    'valid_labels': valid_labels,
    'test_dataset': test_dataset,
    'test_labels': test_labels,
  }
  pickle.dump(save, f, pickle.HIGHEST_PROTOCOL)
  f.close()
except Exception as e:
  print('Unable to save data to', pickle_file, ':', e)
  raise
```

以下是生成 noMNIST.pickle 文件的完整代码：

```python
from __future__ import print_function
import numpy as np
import os
from scipy import ndimage
from six.moves import cPickle as pickle

data_root = '.' # Change me to store data elsewhere

num_classes = 10
np.random.seed(133)

test_folders = ['./notMNIST_small/A', './notMNIST_small/B',
'./notMNIST_small/C', './notMNIST_small/D',
                './notMNIST_small/E', './notMNIST_small/F',
'./notMNIST_small/G', './notMNIST_small/H',
                './notMNIST_small/I', './notMNIST_small/J']
train_folders = ['./notMNIST_large_v2/A', './notMNIST_large_v2/B',
'./notMNIST_large_v2/C', './notMNIST_large_v2/D',
                './notMNIST_large_v2/E', './notMNIST_large_v2/F',
'./notMNIST_large_v2/G', './notMNIST_large_v2/H',
                './notMNIST_large_v2/I', './notMNIST_large_v2/J']

image_size = 28  # Pixel width and height.
pixel_depth = 255.0

def load_letter(folder, min_num_images):
  image_files = os.listdir(folder)
  dataset = np.ndarray(shape=(len(image_files), image_size, image_size),
                       dtype=np.float32)
  num_images = 0
  for image in image_files:
    image_file = os.path.join(folder, image)
    try:
      image_data = (ndimage.imread(image_file).astype(float) -
                    pixel_depth / 2) / pixel_depth
      if image_data.shape != (image_size, image_size):
        raise Exception('Unexpected image shape: %s' %
str(image_data.shape))
      dataset[num_images, :, :] = image_data
      num_images = num_images + 1
    except IOError as e:
      print('Could not read:', image_file, ':', e, '- it\'s ok, skipping.
  dataset = dataset[0:num_images, :, :]
  if num_images < min_num_images:
    raise Exception('Fewer images than expected: %d < %d' %
                    (num_images, min_num_images))
  print('Dataset tensor:', dataset.shape)
  print('Mean:', np.mean(dataset))
  print('Standard deviation:', np.std(dataset))
  return dataset
def maybe_pickle(data_folders, min_num_images_per_class, force=False):
```

```
    dataset_names = []
    for folder in data_folders:
      set_filename = folder + '.pickle'
      dataset_names.append(set_filename)
      if os.path.exists(set_filename) and not force:
        print('%s already present - Skipping pickling.' % set_filename)
      else:
        print('Pickling %s.' % set_filename)
        dataset = load_letter(folder, min_num_images_per_class)
        try:
          with open(set_filename, 'wb') as f:
            #pickle.dump(dataset, f, pickle.HIGHEST_PROTOCOL)
            print(pickle.HIGHEST_PROTOCOL)
            pickle.dump(dataset, f, 2)
        except Exception as e:
          print('Unable to save data to', set_filename, ':', e)
    return dataset_names

def make_arrays(nb_rows, img_size):
    if nb_rows:
      dataset = np.ndarray((nb_rows, img_size, img_size), dtype=np.float32)
      labels = np.ndarray(nb_rows, dtype=np.int32)
    else:
      dataset, labels = None, None
```

让我们来看看之前创建的 pickle 文件如何加载数据并运行一个只含一层隐藏层的网络。

首先，从 noMNIST.pickle 文件中读取训练集、测试集和验证集（ndarray）。

```
with open(pickle_file, 'rb') as f:
 save = pickle.load(f)
 training_dataset = save['train_dataset']
 training_labels = save['train_labels']
 validation_dataset = save['valid_dataset']
 validation_labels = save['valid_labels']
 test_dataset = save['test_dataset']
 test_labels = save['test_labels']

print 'Training set', training_dataset.shape, training_labels.shape
print 'Validation set', validation_dataset.shape, validation_labels.shape
print 'Test set', test_dataset.shape, test_labels.shape
```

将看到类似于以下清单的输出：

```
Training set (1000, 28, 28) (1000,)
Validation set (500, 28, 28) (500,)
Test set (500, 28, 28) (500,)
```

接着，使用 reformat 函数将数据集重新格式化为一个二维数组，以便 TensorFlow 更好地处理。

```
def reformat(dataset, labels):
 dataset = dataset.reshape((-1, image_size *
image_size)).astype(np.float32)
 # Map 0 to [1.0, 0.0, 0.0 ...], 1 to [0.0, 1.0, 0.0 ...]
 labels = (np.arange(num_of_labels) == labels[:, None]).astype(np.float32)
 return dataset, labels
train_dataset, train_labels = reformat(training_dataset, training_labels)
 valid_dataset, valid_labels = reformat(validation_dataset,
validation_labels)
 test_dataset, test_labels = reformat(test_dataset, test_labels)

 print 'Training dataset shape', train_dataset.shape, train_labels.shape
 print 'Validation dataset shape', valid_dataset.shape, valid_labels.shape
 print 'Test dataset shape', test_dataset.shape, test_labels.shape
```

输出结果如下：

```
Training dataset shape (1000, 784) (1000, 10)
Validation dataset shape (500, 784) (500, 10)
Test dataset shape (500, 784) (500, 10)
```

然后，定义一个图，所有的变量都会加载到这个图中。

每一层的权重和偏置在这里给出，其中 image_size = 28, no_of_neurons = 1024。

 应该选取最佳的隐藏层中的神经元数量。过少的神经元会导致较低的准确性，而过多的神经元会导致过拟合。

神经网络中的层	权重	偏置
1	行 =28 × 28=784 列 =1024	1024
2	行 =1024 列 =10	10

我们将初始化 TensorFlow 图并使用训练集、验证集、测试集和标签初始化占位符。我们还将定义两个层的权重和偏置：

```
graph = tf.Graph()
no_of_neurons = 1024
with graph.as_default():
    # Placeholder that will be fed
    # at run time with a training minibatch in the session
    tf_train_dataset = tf.placeholder(tf.float32,
        shape=(batch_size, image_size * image_size))
    tf_train_labels = tf.placeholder(tf.float32, shape=(batch_size,
num_of_labels))
    tf_valid_dataset = tf.constant(valid_dataset)
    tf_test_dataset = tf.constant(test_dataset)

    # Variables.
    w1 = tf.Variable(tf.truncated_normal([image_size * image_size,
no_of_neurons]))
    b1 = tf.Variable(tf.zeros([no_of_neurons]))

    w2 = tf.Variable(
    tf.truncated_normal([no_of_neurons, num_of_labels]))
    b2 = tf.Variable(tf.zeros([num_of_labels]))
```

接下来，定义隐藏层张量并计算 `logits`：

```
hidden1 = tf.nn.relu(tf.matmul(tf_train_dataset, w1) + b1)
logits = tf.matmul(hidden1, w2) + b2
```

我们将使用基于 softmax 函数的交叉熵作为损失函数：

```
loss = tf.reduce_mean(
    tf.nn.softmax_cross_entropy_with_logits(logits=logits,
labels=tf_train_labels))
# Training computation.
loss = tf.reduce_mean(
    tf.nn.softmax_cross_entropy_with_logits(logits=logits,
labels=tf_train_labels))

    # Optimizer.
    optimizer = tf.train.GradientDescentOptimizer(0.5).minimize(loss)
```

接下来，计算 `logits`（预测值）；注意，我们将使用 softmax 函数对 `logits` 进行归一化。

```
train_prediction = tf.nn.softmax(logits)
```

对测试集和验证集进行预测。注意，在这里使用 RELU 激活函数计算 `w1` 和 `b1`。

```
tf.nn.relu(tf.matmul(tf_valid_dataset, w1) + b1)
 valid_prediction = tf.nn.softmax(
    tf.matmul( tf.nn.relu(tf.matmul(tf_valid_dataset, w1) + b1),
             w2
             ) + b2)
 test_prediction = tf.nn.softmax(
    tf.matmul(tf.nn.relu(tf.matmul(tf_test_dataset, w1) + b1), w2) + b2)
```

现在我们将创建一个TensorFlow会话,然后将加载的数据集传入创建的神经网络。

```
with tf.Session(graph=graph) as session:
  tf.initialize_all_variables().run()
  print("Initialized")
  for step in xrange(num_steps):
    offset = (step * batch_size) % (train_labels.shape[0] - batch_size)
    # Generate a minibatch.
    batch_data = train_dataset[offset:(offset + batch_size), :]
    batch_labels = train_labels[offset:(offset + batch_size), :]
    feed_dict = {tf_train_dataset: batch_data, tf_train_labels: batch_labels}
    _, l, predictions = session.run(
      [optimizer, loss, train_prediction], feed_dict=feed_dict)
    minibatch_accuracy = accuracy(predictions, batch_labels)
    validation_accuracy = accuracy(
      valid_prediction.eval(), valid_labels)
    if (step % 10 == 0):
      print("Minibatch loss at step", step, ":", l)
      print("Minibatch accuracy: %.1f%%" % accuracy(predictions, batch_labels))
      print("Validation accuracy: %.1f%%" % validation_accuracy)
minibatch_acc.append( minibatch_accuracy)
validation_acc.append( validation_accuracy)
t = [np.array(minibatch_acc)]
t.append(validation_acc)
```

完整的代码可在https://github.com/rajdeepd/neuralnetwork-programming/blob/ed1/ch02/nomnist/singlelayer-neural_network.py上获取。

完整的代码可以在上述GitHub链接中找到。注意要将验证集的准确率和minibatch的准确率放到我们将绘制的数组中:

```
 print("Test accuracy: %.1f%%" % accuracy(test_prediction.eval(), test_labels))
 title = "NotMNIST DataSet - Single Hidden Layer - 1024 neurons Activation
```

```
function: RELU"
 label = ['Minibatch Accuracy', 'Validation Accuracy']
 draw_plot(x, t, title, label)
```

让我们看一下之前代码绘制的图片:

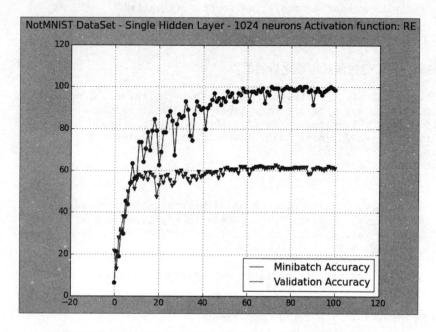

minibatch 的准确率在第 8 次迭代时达到了 100%，而验证集的准确率停留在 60%。

分析激活函数对前馈网络准确率的影响

在前面的例子中，我们使用 RELU 作为激活函数。TensorFlow 支持多种激活函数。让我们看看各个激活函数对验证集准确率的影响。我们将生成一些随机值：

```
x_val = np.linspace(start=-10., stop=10., num=1000)
```

然后产生经过激活函数之后的输出：

```
# ReLU activation
 y_relu = session.run(tf.nn.relu(x_val))
 # ReLU-6 activation
 y_relu6 = session.run(tf.nn.relu6(x_val))
 # Sigmoid activation
 y_sigmoid = session.run(tf.nn.sigmoid(x_val))
```

```
# Hyper Tangent activation
y_tanh = session.run(tf.nn.tanh(x_val))
# Softsign activation
y_softsign = session.run(tf.nn.softsign(x_val))

# Softplus activation
y_softplus = session.run(tf.nn.softplus(x_val))
# Exponential linear activation
y_elu = session.run(tf.nn.elu(x_val))
```

绘制 x_val 经过激活函数之后的图：

```
plt.plot(x_val, y_softplus, 'r--', label='Softplus', linewidth=2)
plt.plot(x_val, y_relu, 'b:', label='RELU', linewidth=2)
plt.plot(x_val, y_relu6, 'g-.', label='RELU6', linewidth=2)
plt.plot(x_val, y_elu, 'k-', label='ELU', linewidth=1)
plt.ylim([-1.5,7])
plt.legend(loc='top left')
plt.title('Activation functions', y=1.05)
plt.show()
plt.plot(x_val, y_sigmoid, 'r--', label='Sigmoid', linewidth=2)
plt.plot(x_val, y_tanh, 'b:', label='tanh', linewidth=2)
plt.plot(x_val, y_softsign, 'g-.', label='Softsign', linewidth=2)
plt.ylim([-1.5,1.5])
plt.legend(loc='top left')
plt.title('Activation functions with Vanishing Gradient', y=1.05)
plt.show()
```

如下图所示：

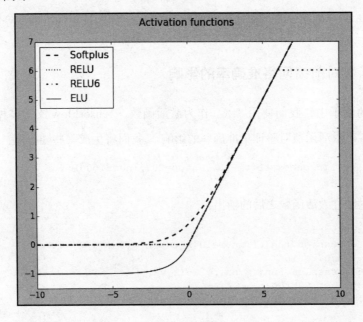

比较不同激活函数和梯度消失的图如下所示：

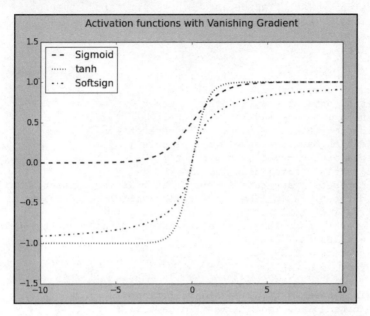

现在让我们看看激活函数以及它们如何影响 NotMNIST 数据验证集的准确率。

我们修改了前面的例子，以便将激活函数作为参数传入 main() 中：

```
RELU = 'RELU'
RELU6 = 'RELU6'
CRELU = 'CRELU'
SIGMOID = 'SIGMOID'
ELU = 'ELU'
SOFTPLUS = 'SOFTPLUS'
def activation(name, features):
  if name == RELU:
    return tf.nn.relu(features)
  if name == RELU6:
    return tf.nn.relu6(features)
  if name == SIGMOID:
    return tf.nn.sigmoid(features)
  if name == CRELU:
    return tf.nn.crelu(features)
  if name == ELU:
    return tf.nn.elu(features)
  if name == SOFTPLUS:
    return tf.nn.softplus(features)
```

run() 函数的定义包含了之前定义的各种信息。

```
batch_size = 128
#activations = [RELU, RELU6, SIGMOID, CRELU, ELU, SOFTPLUS]
activations = [RELU, RELU6, SIGMOID, ELU, SOFTPLUS]
plot_loss = False
def run(name):
 print(name)
 with open(pickle_file, 'rb') as f:
    save = pickle.load(f)
    training_dataset = save['train_dataset']
    training_labels = save['train_labels']
    validation_dataset = save['valid_dataset']
    validation_labels = save['valid_labels']
    test_dataset = save['test_dataset']
    test_labels = save['test_labels']
 train_dataset, train_labels = reformat(training_dataset, training_labels)
 valid_dataset, valid_labels = reformat(validation_dataset,
     validation_labels)
 test_dataset, test_labels = reformat(test_dataset, test_labels)
 graph = tf.Graph()
 no_of_neurons = 1024
 with graph.as_default():

    tf_train_dataset = tf.placeholder(tf.float32,
    shape=(batch_size, image_size * image_size))
    tf_train_labels = tf.placeholder(tf.float32, shape=(batch_size,
        num_of_labels))
    tf_valid_dataset = tf.constant(valid_dataset)
    tf_test_dataset = tf.constant(test_dataset)
    # Define Variables.
    # Training computation...
    # Optimizer ..
    # Predictions for the training, validation, and test data.
    train_prediction = tf.nn.softmax(logits)
    valid_prediction = tf.nn.softmax(
    tf.matmul(activation(name,tf.matmul(tf_valid_dataset, w1) + b1), w2) + b2)
    test_prediction = tf.nn.softmax(
    tf.matmul(activation(name,tf.matmul(tf_test_dataset, w1) + b1), w2) + b2)

 num_steps = 101
 minibatch_acc = []
 validation_acc = []
 loss_array = []
 with tf.Session(graph=graph) as session:
    tf.initialize_all_variables().run()
    print("Initialized")
    for step in xrange(num_steps):
        offset = (step * batch_size) % (train_labels.shape[0] - batch_size)
        # Generate a minibatch.
        batch_data = train_dataset[offset:(offset + batch_size), :]
        batch_labels = train_labels[offset:(offset + batch_size), :]
        feed_dict = {tf_train_dataset: batch_data, tf_train_labels:
batch_labels}
```

```
    _, l, predictions = session.run(
     [optimizer, loss, train_prediction], feed_dict=feed_dict)
    minibatch_accuracy = accuracy(predictions, batch_labels)
    validation_accuracy = accuracy(
     valid_prediction.eval(), valid_labels)
    if (step % 10 == 0):
      print("Minibatch loss at step", step, ":", l)
      print("Minibatch accuracy: %.1f%%" % accuracy(predictions,
       batch_labels))
      print("Validation accuracy: %.1f%%" % accuracy(
    valid_prediction.eval(), valid_labels))
    minibatch_acc.append(minibatch_accuracy)
    validation_acc.append(validation_accuracy)
    loss_array.append(l)
    print("Test accuracy: %.1f%%" % accuracy(test_prediction.eval(),
   test_labels))
  return validation_acc, loss_array
```

结果如图 2-2 所示：

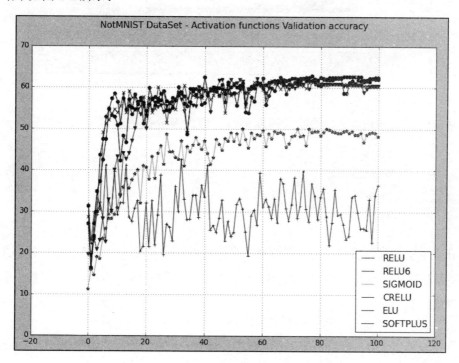

图 2-2　各种激活函数的验证集准确率图示

从图 2-2 中可以看出，RELU 和 RELU6 在验证集上的准确率最高，接近 60%。现在让我们来看看采用各种激活函数时，训练集上的损失是多少（见图 2-3）。

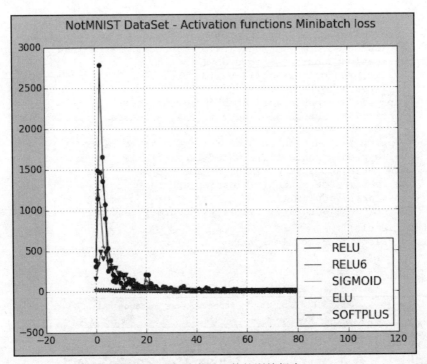

图 2-3　各种激活函数的训练损失

尽管在短时间内 RELU 的收敛速度最慢，但大部分激活函数的训练损失都会收敛到零。

2.6　总结

在本章中，我们构建了第一个神经网络，这个网络只是一个前馈网络。我们用它对 Iris 数据集和之后的 NotMNIST 数据集进行分类。你已经了解了各种激活函数对验证集准确率的影响。

在下一章中，我们将探索卷积神经网络，这种网络对于图像数据集更加先进和有效。

CHAPTER 3
第 3 章

神经网络的优化

深度学习中的很多应用需要优化。优化是指把我们正在处理的任何问题推向最终的状态。我们必须为优化过程提供数据，这些数据为函数提供模型的常量和参数，并描述在某些约束条件下的总体目标。

在本章中，我们将看一下 TensorFlow 库提供的 TensorFlow 处理流程和不同的优化模型。主要内容如下：

- ❑ 优化的基础。
- ❑ 优化器的类型。
- ❑ 梯度下降。
- ❑ 选择正确的优化器。

3.1 什么是优化

优化是基于约束找到最大值或最小值的过程。深度学习模型优化算法的选择可能意味着在模型取得较好效果的时间成本上有几分钟、几小时、几天的差别。

优化是深度学习的核心。大部分学习问题都归结为优化问题。假设我们正针对一些数据集解决一个问题，使用这个预处理过的数据，我们通过求解优化问题来训练一个模型，求解优化问题就是在考虑所选损失函数和正则化函数的情况下，使模型权重最优。

模型的超参数在模型的充分训练中起非常大的作用。因此，使用不同的优化策略和算法来评估模型的超参数是否正确且最优至关重要，这将影响模型的学习过程，最终影响模型的输出。

3.2 优化器的类型

首先，我们先看看高阶优化算法，然后深入了解各个优化器。

一阶优化算法使用相关参数的梯度值来使损失函数最小化或最大化。常用的一阶优化算法是梯度下降法。一阶导数告诉我们函数在特定点上是递减的还是递增的。一阶导数为我们提供一条与误差表面相切的直线。

函数的导数取决于单个变量，而函数的梯度取决于多个变量。

二阶优化算法使用二阶导数（也称为 Hessian）来最小化或最大化给定的损失函数。在这，Hessian 是二阶偏导数矩阵。计算二阶导数的成本很高，因此并不常用。二阶导数表明一阶导数是增加还是减小的，给出了函数曲率的概念。二阶导数为我们提供了一个能触及误差曲面形状的二次曲面。

二阶导数计算成本很高，但是二阶优化方法的优点是它不忽略曲面的曲率。而且随着迭代次数的增加，效果更好。选择优化器需要注意的一点是，一阶优化方法计算简单耗时少，在大型数据集上收敛速度较快。二阶优化方法只有在二阶导数已知的情况下才会更快，而且二阶优化方法在同样的时间和内存资源的情况下，计算成本较高，速度较慢。

二阶优化方法有时候可以比一阶优化的梯度下降方法表现更好。因为二阶优化方法永远不会卡在慢收敛路径附近，即鞍点附近；而梯度下降有时会卡住不收敛。

3.3 梯度下降

梯度下降是使函数最小化的算法。一组参数定义一个函数，梯度下降算法从初始的参数值开始，并迭代地向使函数的参数最小化的方向移动。

使用微积分，在函数梯度下降的方向上逐步迭代以实现损失函数最小化，如图 3-1 所示。

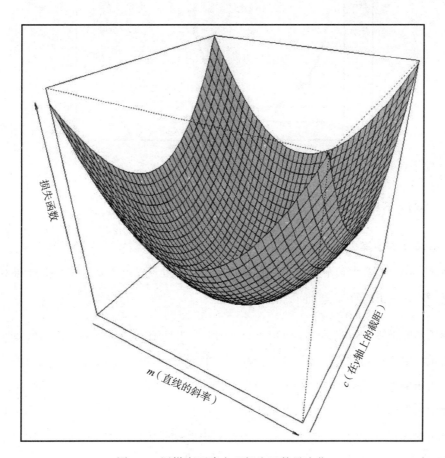

图 3-1 用梯度下降实现损失函数最小化

梯度下降是最为成功的优化算法。如前所述，它用于在神经网络中更新权重，以使损失函数最小化。现在，我们来谈谈一个重要的神经网络方法——反向传播。首先，我们要做前向传播，计算输入节点和相应权重的点积，然后将对应点积相加，对它们使用激活函数，对输入和输出进行转换，并在模型上添加非线性变换。这样会使模型学习几乎任意的函数映射。

之后，在神经网络中做反向传播，携带误差项，使用梯度下降更新权重值，如图 3-2 所示。

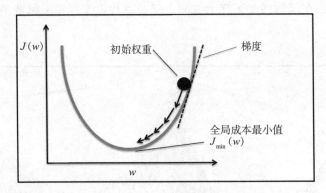

图 3-2　反向传播示意图

3.3.1　梯度下降的变体

标准的梯度下降也称为**批量梯度下降**。它计算整个数据集的梯度，但只执行一次更新。因此，对于非常大且不适合内存装载的数据集，批量梯度下降是相当缓慢和困难的。现在我们来看一下能解决这个问题的优化算法。

随机梯度下降（Stochastic Gradient Descent，SGD）在每个训练样例上执行参数更新，而小批量梯度下降是在每个批次中对 n 个训练样例执行一次更新。随机梯度下降的问题在于，由于频繁地更新参数和梯度的波动，它最终会使得（损失）收敛到正确的最小值这一过程变得复杂，并且因为频繁的波动，导致溢出。小批量梯度下降解决了这个问题，在参数更新的过程中降低了方差，使收敛过程更好更稳定。SGD 和小批量梯度下降可交换使用。

梯度下降的整体问题包括选择合适的学习率以避免在较小值处收敛缓慢或较大值处的收敛异常；把相同的学习率应用于所有参数的更新，其中如果数据是稀疏的，我们可能不想把所有权重更新至相同的程度；最后，处理鞍点。

3.3.2 优化梯度下降的算法

下面介绍各种优化梯度下降的方法，以现实为每个参数计算不同的学习率、计算动量、减缓学习率的衰减。

为了解决 SGD 方差高扰动的问题，我们设计了一种称为**动量**的方法。该方法通过沿着梯度的方向进行导向和削弱无关梯度方向来加速 SGD。基本上，该方法将上一步的向量更新信息添加至当前向量更新，动量值通常设定为 0.9，动量方法使收敛更快、更稳定，减少震荡。

牛顿加速梯度（NAG）方法阐述了当我们达到最小值（也就是曲线上的最低点）时，动量相当高，因此并不知道在该处如何放缓速度，这会导致错过全局最小值并继续移动。牛顿法提出，先基于前面的动量做大幅度的跳跃，然后计算梯度并对参数进行修正（对参数做一次更新）。现在，这种更新方法能防止更新过快而错过最小值，对变化反应更敏感。

自适应梯度（Adagrad）方法允许学习率根据参数而调整。因此，它对低频参数做较大幅度的更新，对高频参数做较小幅度的更新。所以，它非常适合处理稀疏数据。该方法的主要缺陷是学习率总在下降和衰减。学习率下降的问题可以通过 Adadelta 方法来解决。

自适应学习率调整（Adadelta）方法解决了 Adagrad 方法中学习率下降的问题。在 Adagrad 方法中，学习率通过 1 除平方根之和来计算。在每个阶段，加上一个平方根并求和，这会导致学习率不断变小。Adadelta 方法不是对之前的所有平方根做累加，而是使用滑动窗口的方式，这种方法允许总和减少。

自适应动量估计（Adam）方法计算每个参数的自适应学习率，像 Adadelta 方法一样，Adam 方法不仅存储之前平方梯度的衰减平均值，而且存储每个参数的动量变化。Adam 方法在实践中表现很好，是当今最常用的优化方法之一。

图 3-3 和图 3-4（图片来源：Alec Radford）展示了前面介绍的优化算法的优化效果。我们观察各种优化算法在损失曲面上随着时间推移的表现。Adagrad、RMSProp 和 Adadelta 几乎快速朝着正确的下降方向前进，并快速收敛。而动量和 NAG 则偏离了方

向。由于 NAG 会向前更新梯度并且朝着最小点前进,它的收敛速度得到提升,NAG 能够很快地进行修正。

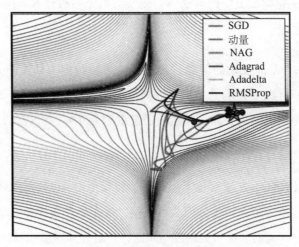

图 3-3　各种优化算法的效果图

图 3-4 展示了不同优化算法在鞍点附近的行为。对于 SGD、**动量**和 NAG 来说,虽然这种对称性很有挑战,但是它们慢慢地设法摆脱了鞍点,而 Adagrad、Adadelta 和 RMSProp 则沿着负斜率的方向下降,如图 3-4 所示。

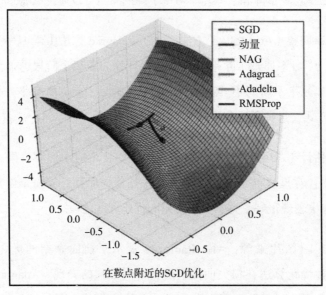

图 3-4　各种优化算法在鞍点附近的行为图示

3.4 优化器的选择

在输入数据为稀疏数据或者训练复杂神经网络想快速收敛的情况下，使用自适应学习率会取得较好的结果，我们也不需要调整学习率，大多数情况下，Adam 都是一个不错的选择。

优化的例子

以线性回归为例，我们通过最小化直线到每个点的距离平方和，来找到通过多个数据点的直线的最佳拟合。这就是为什么称之为最小二乘回归。本质上，我看把这个问题看作一个优化问题，使损失函数最小化。

让我们设置输入并看一下散点图：

```
# input data
xData = np.arange(100, step=.1)
yData = xData + 20 * np.sin(xData/10)
```

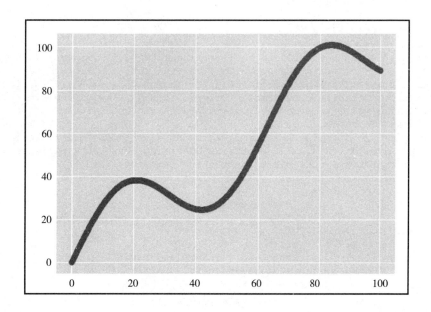

定义数据集大小和批次大小：

```
# define the data size and batch size
nSamples = 1000
batchSize = 100
```

我们需要调整数据集的大小以符合 TensorFlow 的输入格式，如下所示：

```
# resize input for tensorflow
 xData = np.reshape(xData, (nSamples, 1))
 yData = np.reshape(yData, (nSamples, 1))
```

下面的代码对权重（weights）和偏置（bias）进行初始化，并给出线性模型和损失函数。

```
with tf.variable_scope("linear-regression-pipeline"):
    W = tf.get_variable("weights", (1,1),
initializer=tf.random_normal_initializer())
    b = tf.get_variable("bias", (1, ),
initializer=tf.constant_initializer(0.0))

    # model
    yPred = tf.matmul(X, W) + b
    # loss function
    loss = tf.reduce_sum((y - yPred)**2/nSamples)
```

然后，通过设定优化器使损失最小化：

```
# set the optimizer
 #optimizer = tf.train.GradientDescentOptimizer(learning_rate=0.001).minimize(loss)
 #optimizer = tf.train.AdamOptimizer(learning_rate=.001).minimize(loss)
 #optimizer = tf.train.AdadeltaOptimizer(learning_rate=.001).minimize(loss)
 #optimizer = tf.train.AdagradOptimizer(learning_rate=.001).minimize(loss)
 #optimizer = tf.train.MomentumOptimizer(learning_rate=.001, momentum=0.9).minimize(loss)
 #optimizer = tf.train.FtrlOptimizer(learning_rate=.001).minimize(loss)
 optimizer = tf.train.RMSPropOptimizer(learning_rate=.001).minimize(loss)
We then select the mini batch and run the optimizers errors = []
with tf.Session() as sess:
    # init variables
    sess.run(tf.global_variables_initializer())

    for _ in range(1000):
        # select mini batch
        indices = np.random.choice(nSamples, batchSize)
        xBatch, yBatch = xData[indices], yData[indices]
        # run optimizer
        _, lossVal = sess.run([optimizer, loss], feed_dict={X: xBatch, y: yBatch})
        errors.append(lossVal)
```

```
plt.plot([np.mean(errors[i-50:i]) for i in range(len(errors))])
plt.show()
plt.savefig("errors.png")
```

上述代码的输出结果如下所示：

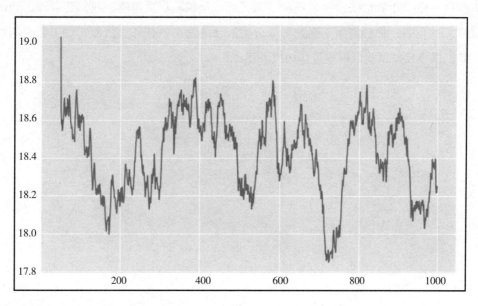

通过滑动得到滑动曲线，如下所示：

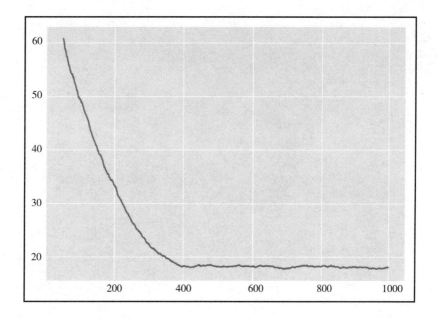

3.5 总结

在本章，我们学习了优化技术的基础和不同类型的优化器。优化是一个复杂的主题，很大程度上取决于数据的性质和规模。而且，优化依赖权重矩阵。我们会训练、调优很多这样的优化器，将它们用于诸如图像分类或预测之类的任务。然而，对于新的场景应用，我们需要反复试验以确定最佳解决方案。

CHAPTER 4

第 4 章

卷积神经网络

卷积网络（参考 LeCun [1]，2013）也称为**卷积神经网络**或 **CNN**，是一种用类似网格拓扑处理数据的特殊类型的神经网络。例如，时间序列的数据可以看作是一维网络以固定时间间隔对数据进行采样的结果，图像数据可以看作是二维的像素点。卷积神经网络这一名称意味着网络采用称为**卷积**的数学运算。卷积是一种特殊的线性操作。卷积网络是使用卷积（数学运算）代替其至少一个层中的一般矩阵乘法的神经网络。

首先，我们将描述卷积的数学运算。然后讨论池化的概念以及它在 CNN 中的作用。我们也将看看 TensorFlow 中的卷积网络实现。

在本章的最后，我们将使用 TensorFlow 的 CNN 实现对斯坦福数据集中的狗和猫进行分类。

Lecun[1]:http://yann.lecun.com/exdb/lenet/。

本章将涉及以下主题：

- 卷积神经网络概述和直观理解。
- 卷积操作。
- 池化。
- 用卷积网络进行图像分类。

4.1 卷积神经网络概述和直观理解

CNN 由多层卷积层、池化层和最后的全连接层组成。这种网络结构比我们在第 2 章中讨论的纯前馈网络有效率的多。

如图 4-1 所示,图片通过**卷积层→最大池化层→卷积层→最大池化层→全连接层**。这就是卷积神经网络的结构。

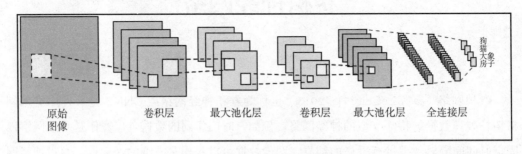

图 4-1　卷积神经网络结构图示

4.1.1 单个卷积层的计算

我们先来直观地看一下卷积层的计算。卷积层的参数由一组可学习的卷积核(也称为**张量**)组成。每个卷积核的大小(深度、宽度和高度)都很小,但是会延伸到输入数据(图像)的整个深度。卷积网络第一层的卷积核通常具有 5×5×3 的大小(即,高度和宽度为 5 个像素,深度为 3 个像素,因为图像具有 3 个颜色通道的深度)。在前向传播过程中,卷积核在输入的宽度和高度上进行滑动(或**卷积**),并在每一个点计算卷积核和输入之间的点积。当卷积核在输入的宽度和高度上滑动时,会产生一个二维的激活图,这个激活图给出了该卷积核在每个输入点上的响应。网络将学习那些对某种视觉特征(例如某个方向的边缘或第一层上的某种颜色的斑点)产生激活的卷积核,在更高层的网络上也将学习那些可以发现整个结构或者轮廓的卷积核。一旦我们在每个卷积层中都有一整套卷积核(例如 12 个卷积核),每个卷积核都会生成一个单独的二维激活图。我们沿着深度堆叠这些激活图并产生输出。

图 4-2 展示了用 5×5×3 的卷积核对 32×32×3 的图像进行卷积,结果如图 4-3 所示。

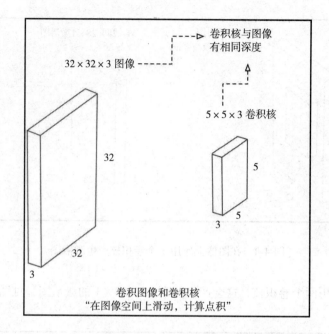

图 4-2 用 5×5 卷积核对 32×32 像素的图像进行卷积

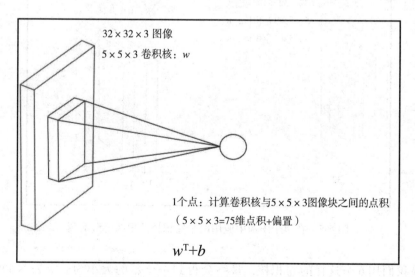

图 4-3 在一个点上计算卷积核与图像块的点积

接着,让我们用先前的卷积核对整个图像进行卷积,每次移动一个像素。

最后输出大小为 $28\times28\times1$ 的图,称为**激活图**,见图 4-4。

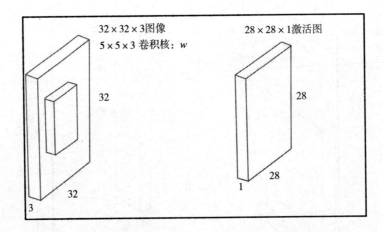

图 4-4 在图像上应用 1 个卷积核产生激活图

考虑逐个使用两个卷积核,这会产生两个 28×28×1 的激活图,见图 4-5。

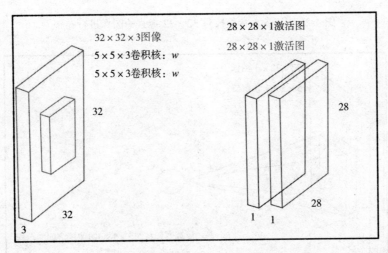

图 4-5 在单个图像中应用两个卷积核产生两个激活图

如果我们用 6 个这样的卷积核,最终会得到一个新的大小为 28×28×6 的图,见图 4-6。卷积网络就是由一系列这样的卷积层所组成的,其中穿插着诸如 Relu 等激活函数。

让我们根据 TensorFlow 的用语正式定义 CNN。

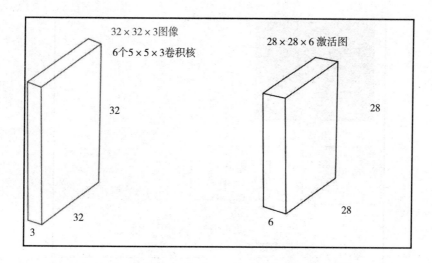

图 4-6　应用 6 个 $5 \times 5 \times 3$ 的卷积核生成 $28 \times 28 \times 6$ 的激活图

定义：CNN 是一个在输入和输出之间至少有一层（`tf.nn.conv2d`）由可学习的卷积核产生输出的神经网络。使用卷积核对输入层（张量）的每个点进行卷积。通过在输入张量上滑动（卷积）来生成一个卷积后的输出。

应用例子：图 4-7 是通过卷积操作在输入图像上使用边缘检测卷积核的例子。

CNN 匹配信息的过程和在猫的初级视皮层的细胞中发现的结构相似。当信号通过猫的初级视皮层时，某些层会对一些突出的视觉模式产生信号。例如，当一条水平线穿过它时，某一层的单元将激活（增加其输出信号）。CNN 将表现出类似的行为，其中神经元集群根据从训练中学习到的模式激活。在经过预先标注的数据训练之后，当一条水平或垂直的线经过网络时，CNN 中的某些层将会激活。

可以匹配水平或者垂直的线是有用的神经网络结构，但是 CNN 将多个简单模式组成一层以此对复杂模式进行匹配。这些模式称为**滤波器**或者**卷积核**。训练的目的是调整卷积核的权重来最小化损失函数（的值）。可以将多层组合到一起然后通过梯度下降或者其他优化方法训练卷积核参数。

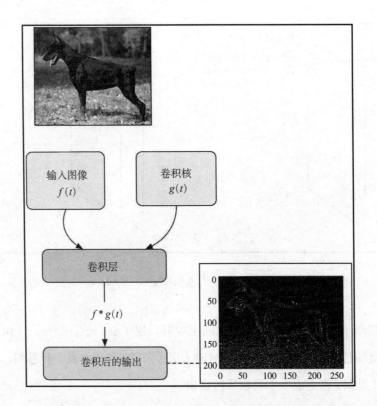

图 4-7 在输入图像上应用卷积核的边缘检测

4.1.2 TensorFlow 中的 CNN

CNN 由卷积层(由 `tf.nn.conv2d` 定义)、非线性层(`tf.nn.relu`)、最大池化层(`tf.nn.max_pool`)和全连接层(`tf.matmul`)组成。图 4-8 显示了典型 CNN 层及其在 TensorFlow 中相应的实现。

TensorFlow 中读取图片

现在让我们看看 TensorFlow 是如何读取图片的。首先定义一个包含 3 张图片的常量数组,然后将它们读取到一个会话中。

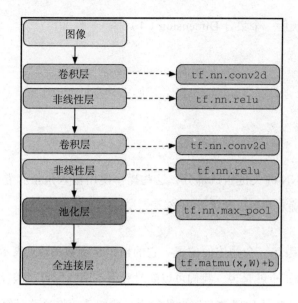

图 4-8 CNN 层到 TensorFlow 函数的映射

```
sess = tf.InteractiveSession()
image_batch = tf.constant([
    [ # First Image
      [[255, 0, 0], [255, 0, 0], [0, 255, 0]],
      [[255, 0, 0], [255, 0, 0], [0, 255, 0]]
    ],
    [ # Second Image
      [[0, 0, 0], [0, 255, 0], [0, 255, 0]],
      [[0, 0, 0], [0, 255, 0], [0, 255, 0]]
    ],
    [ # Third Image
      [[0, 0, 255], [0, 0, 255], [0, 0, 255]],
      [[0, 0, 255], [0, 0, 255], [0, 0, 255]]
    ]
])
print(image_batch.get_shape())
print(sess.run(image_batch)[1][0][0])
```

之前的输出是张量的大小以及第一张图片的第一个像素值。在本例中,图片数组包含 3 张图片。每张图片有 RGB 3 个通道,每个通道的高度为 2 个像素,宽度为 3 个像素。示例代码会输出图片的数量,即第一个集合 Dimension(1)的大小。图片的高度是第二个集合 Dimension(2)的大小。图片的宽度是第三个集合 Dimension(3)的大小。

图片通道的大小是最后一个集合 Dimension（4）。

```
(3, 2, 3, 3)
[255 0 0]
```

4.2 卷积操作

卷积操作是 CNN 的关键组成部分；这些操作使用输入张量和卷积核来计算输出。卷积操作的关键是确定参数以调优卷积核。

假设我们正在跟踪一个物体的位置。它的输出是函数 $x(t)$，表示物体在时间 t 的位置。x 和 t 都是实数，也就是说，我们可以随时获得不同的位置。假设我们计算出的值有噪声。为了使计算的物体位置含有较少的噪声，我们希望将这些计算值做一个平均。时间越近的计算值对我们来说更为重要。我们希望这是一个给越近的计算值更高的权重的加权平均。我们可以使用加权函数 $w(a)$ 来计算加权平均，其中 a 是计算的时间（进行计算时）。如果我们在每个时刻应用加权平均运算，会得到一个新的、对物体位置的估计经过了平滑的函数：

$$s(t) = \int x(a)w(t-a)$$

这个操作叫作**卷积**。卷积操作用星号表示：

$$s(t) = (x*w)(t)$$

这里，

- w 是卷积核。
- x 是输入。
- s 是输出，也称为**特征图**。

4.2.1 对图像进行卷积

如果我们将二维图像作为输入,那么我们的卷积核也是二维的。之前的公式变为如下形式:

$$s(i,j)=(I*K)(i,j)=\sum_{m}\sum_{n}I(m,n)*K(i-m,j-n)$$

由于卷积函数是可交换的,我们可以把前面的等式写成如下形式:

$$s(i,j)=(I*K)(i,j)=\sum_{m}\sum_{n}I(i-m,j-n)*K(m,n)$$

根据互相关函数的性质,把 $i-m$ 和 $j-n$ 变为 $i+m$ 和 $j+n$,这就是 TensorFlow 中的实现:

$$s(i,j)=(I*K)(i,j)=\sum_{m}\sum_{n}I(i+m,j+n)*K(m,n)$$

我们在 TensorFlow 中定义一个简单的输入和一个卷积核然后执行 `conv2d` 操作。我们观察一个简单的输入图像和卷积核。图 4-9 展示了对一个基本的图像和一个卷积核进行卷积以及卷积后的结果。

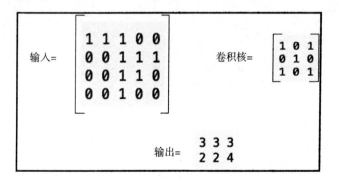

图 4-9 对基本图像与卷积核进行卷积的示例

对于步长为 1,1,1,1 其产生的结果如图 4-10 所示。

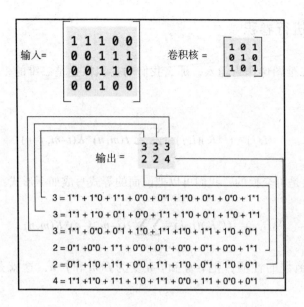

图 4-10　示例输出

接着，我们在 TensorFlow 中实现同样的操作。

```
i = tf.constant([
            [1.0, 1.0, 1.0, 0.0, 0.0],
            [0.0, 0.0, 1.0, 1.0, 1.0],
            [0.0, 0.0, 1.0, 1.0, 0.0],
            [0.0, 0.0, 1.0, 0.0, 0.0]], dtype=tf.float32)
k = tf.constant([
            [1.0, 0.0, 1.0],
            [0.0, 1.0, 0.0],
            [1.0, 0.0, 1.0]
      ], dtype=tf.float32),
kernel = tf.reshape(k, [3, 3, 1, 1], name='kernel')
image = tf.reshape(i, [1, 4, 5, 1], name='image')
res = tf.squeeze(tf.nn.conv2d(image, kernel, strides=[1, 1, 1, 1],
padding="VALID"))
# VALID means no padding
with tf.Session() as sess:
    print sess.run(res)
```

输出结果如下所示，这和我们人工计算的结果是一样的。

```
[[ 3. 3. 3.]
 [ 2. 2. 4.]]
```

4.2.2 步长

卷积操作的主要目的是降低图像的维度（宽度，高度，通道数）。图像越大，需要处理的时间就越长。

strides 参数使得卷积核跳过图像中的一些像素，它们不会包含在输出中。当一幅较大的图像和较复杂的卷积核进行卷积时，strides 参数决定了卷积核是如何进行卷积操作的。卷积就是将卷积核在输入上进行滑动，strides 操作正是决定它是如何进行滑动的，而不是遍历输入的每个元素。

我们看一看下面的例子。在这个例子中，我们用 3×3×1 的卷积核对 6×6×1 的图像进行卷积，步长为 1,3,3,1，见图 4-11。

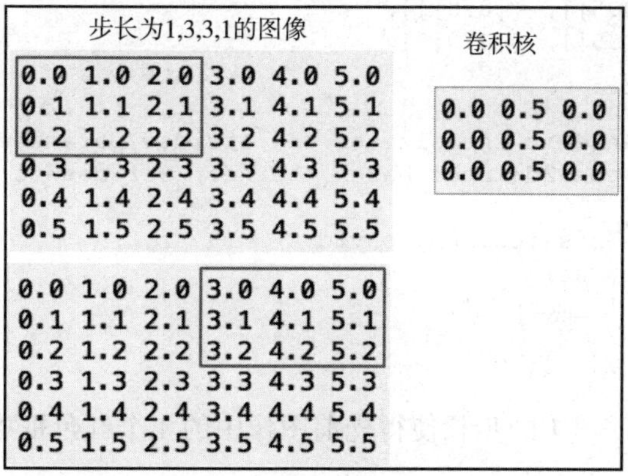

```
步长为1,3,3,1的图像              卷积核
0.0 1.0 2.0  3.0 4.0 5.0
0.1 1.1 2.1  3.1 4.1 5.1      0.0 0.5 0.0
0.2 1.2 2.2  3.2 4.2 5.2      0.0 0.5 0.0
0.3 1.3 2.3  3.3 4.3 5.3      0.0 0.5 0.0
0.4 1.4 2.4  3.4 4.4 5.4
0.5 1.5 2.5  3.5 4.5 5.5


0.0 1.0 2.0  3.0 4.0 5.0
0.1 1.1 2.1  3.1 4.1 5.1
0.2 1.2 2.2  3.2 4.2 5.2
0.3 1.3 2.3  3.3 4.3 5.3
0.4 1.4 2.4  3.4 4.4 5.4
0.5 1.5 2.5  3.5 4.5 5.5
```

图 4-11　第 1 步：用步长为 1,3,3,1 的卷积核进行滑动

在第 3 步和第 4 步，卷积核将和以下元素进行计算，见图 4-12。

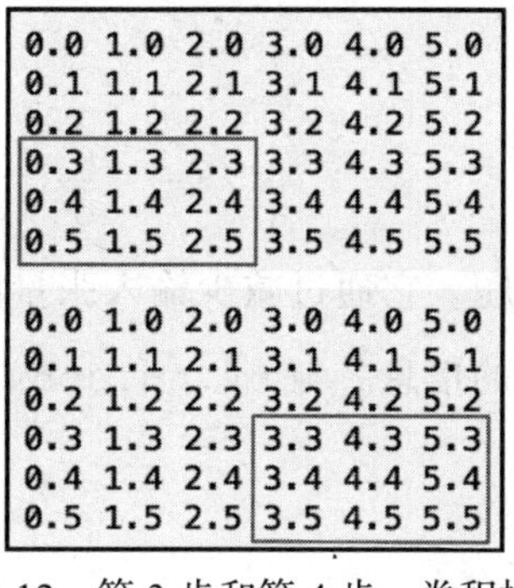

```
0.0 1.0 2.0 3.0 4.0 5.0
0.1 1.1 2.1 3.1 4.1 5.1
0.2 1.2 2.2 3.2 4.2 5.2
0.3 1.3 2.3 3.3 4.3 5.3
0.4 1.4 2.4 3.4 4.4 5.4
0.5 1.5 2.5 3.5 4.5 5.5


0.0 1.0 2.0 3.0 4.0 5.0
0.1 1.1 2.1 3.1 4.1 5.1
0.2 1.2 2.2 3.2 4.2 5.2
0.3 1.3 2.3 3.3 4.3 5.3
0.4 1.4 2.4 3.4 4.4 5.4
0.5 1.5 2.5 3.5 4.5 5.5
```

图 4-12　第 3 步和第 4 步：卷积操作

在 TensorFlow 中实现以上操作。输出是一个 2×2×1 的张量。

```python
import tensorflow as tf

def main():
    session = tf.InteractiveSession()
    input_batch = tf.constant([
        [ # First Input (6x6x1)
            [[0.0], [1.0], [2.0], [3.0], [4.0], [5.0]],
            [[0.1], [1.1], [2.1], [3.1], [4.1], [5.1]],
            [[0.2], [1.2], [2.2], [3.2], [4.2], [5.2]],
            [[0.3], [1.3], [2.3], [3.3], [4.3], [5.3]],
            [[0.4], [1.4], [2.4], [3.4], [4.4], [5.4]],
            [[0.5], [1.5], [2.5], [3.5], [4.5], [5.5]],
        ],
    ])
    kernel = tf.constant([ # Kernel (3x3x1)
        [[[0.0]], [[0.5]], [[0.0]]],
        [[[0.0]], [[0.5]], [[0.0]]],
        [[[0.0]], [[0.5]], [[0.0]]]
    ])
    # NOTE: the change in the size of the strides parameter.
    conv2d = tf.nn.conv2d(input_batch, kernel, strides=[1, 3, 3, 1], padding='SAME')
    conv2d_output = session.run(conv2d)
    print(conv2d_output)
if __name__ == '__main__':
    main()
```

输出如下所示。1,3,3,1 的步长使得先前图像中的 4 个红色框中的像素点和卷积核进行相乘：

```
[[[[ 1.64999998][ 6.1500001 ]]
  [[ 2.0999999 ][ 6.60000038]]]]
```

4.3 池化

池化层可以减少过拟合问题。它通过减少输入张量的大小来提升模型性能。通常，它们用于缩小输入，保留重要的信息。与 `tf.nn.conv2d` 相比，池化是一种更快速的减小输入数据大小的机制。

TensorFlow 支持以下池化机制：

- 均值池化。
- 最大池化(返回最大值所在索引)。
- 随机池化。

每个池化操作都在大小为 `ksize` 的窗口中进行。这些窗口的间隔为 `strides`。如果 `strides` 全为 1 (1, 1, 1, 1),那么所有的窗口都会被使用;如果 `strides` 全为 2 (1, 2, 2, 1),那么在每个维度上,每隔一个窗口会被使用,以此类推。

4.3.1 最大池化

以下定义的函数对输入的四维张量进行最大池化:

```
max_pool(
    value, ksize, strides, padding, data_format='NHWC', name=None
)
```

以上参数的含义如下:

- `value`:进行最大池化的四维张量。大小为 `[batch, height, width, channels]`,类型为 `tf.float32`。
- `ksize`:整型列表,长度大于等于 4。输入张量在每个维度上的窗口大小。
- `strides`:整型列表,长度大于等于 4。输入张量在每个维度上滑动窗口的步长。
- `padding`:`string` 类型,`VALID` 或者 `SAME`。填充算法。图 4-13 解释 `VALID` 和 `SAME` 填充方式。

参考:https://stackoverflow.com/questions/37674306/whatis-the-difference-between-same-and-valid-padding-in-tf-nnmax-pool-of-t。

- `data_format`:`string` 类型,支持 `NHWC` 和 `NCHW`。
- `name`:可选,操作的名称。

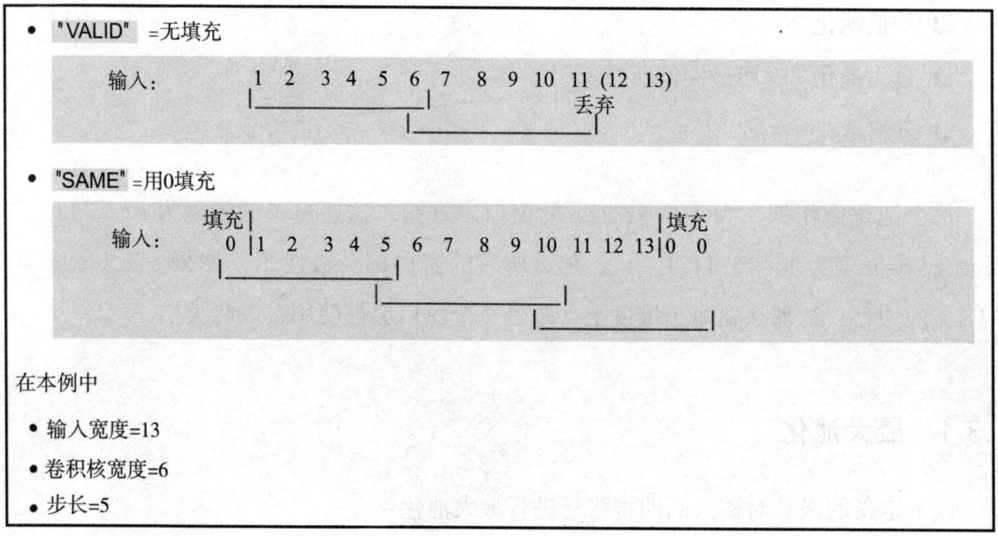

```
• "VALID" =无填充
    输入:      1  2  3  4  5  6  7  8  9  10 11 (12 13)
               |_____|        |_____|    丢弃
                          |_____|

• "SAME" =用0填充
              填充|                                  |填充
    输入:      0 |1  2  3  4  5  6  7  8  9 10 11 12 13| 0  0
               |_____|
                          |_____|
                                    |_____|
```

在本例中

- 输入宽度=13
- 卷积核宽度=6
- 步长=5

图 4-13　填充示例

4.3.2　示例代码

以下代码使用 VALID 填充方式对张量进行最大池化。

```
import tensorflow as tf

batch_size=1
input_height = 3
input_width = 3
input_channels = 1

def main():
    sess = tf.InteractiveSession()
    layer_input = tf.constant([
        [
            [[1.0], [0.2], [2.0]],
            [[0.1], [1.2], [1.4]],
            [[1.1], [0.4], [0.4]]
        ]
    ])

    # The strides will look at the entire input by using the image_height and
    image_width
    kernel = [batch_size, input_height, input_width, input_channels]
    max_pool = tf.nn.max_pool(layer_input, kernel, [1, 1, 1, 1], "VALID")
    print(sess.run(max_pool))
```

```
if __name__ == '__main__':
    main()
```

输出结果为 3×3×1 窗口中的最大值:

```
[[[[ 2.]]]]
```

下图说明了最大值池化的执行逻辑:

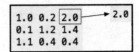

可以看到，在 1,1,1 步长下，最大池化从窗口中选取最大值。

均值池化

它对输入张量做均值池化操作。输出的每个元素都是对应大小为 ksize 窗口中元素的平均值。它使用 tf.nn.avg_pool 方法定义:

```
avg_pool( value, ksize, strides, padding, data_format='NHWC', name=None)
```

我们来看一个在简单的 2D 张量中使用 avg_pool 的代码示例:

```
import tensorflow as tf

batch_size=1
input_height = 3
input_width = 3
input_channels = 1

def main():
    sess = tf.InteractiveSession()
    layer_input = tf.constant([
        [
            [[1.0], [0.2], [2.0]],
            [[0.1], [1.2], [1.4]],
            [[1.1], [0.4], [0.4]]
        ]
    ])

    # The strides will look at the entire input by using the image_height and image_width
```

```
    kernel = [batch_size, input_height, input_width, input_channels]
    avg_pool = tf.nn.avg_pool(layer_input, kernel, [1, 1, 1, 1], "VALID")
    print(sess.run(avg_pool))

if __name__ == '__main__':
    main()
```

输出结果是张量中所有值的平均值。

平均值＝(1.0＋0.2＋2.0＋0.1＋1.2＋1.4＋1.1＋0.4＋0.4)/9＝0.86666

[[[[0.86666667]]]]

4.4 使用卷积网络进行图像分类

现在让我们看一个使用 CNN 网络的更真实的例子。我们会用斯坦福大学的猫和狗的数据集。这个数据集有 100 多张猫和狗的图片。

你可以从下面的地址下载数据集（猫和狗各 100 张图）：https://s3.amazonaws.com/neural-networking-book/ch04/dogs_vs_cats.tar.gz。

1）导入相关函数和 Python 类。

```
import matplotlib.pyplot as plt
import tensorflow as tf
import pandas as pd
import numpy as np
from sklearn.metrics import confusion_matrix
import time
from datetime import timedelta
import math
import dataset
import random
```

2）我们定义卷积层的参数。有 3 个卷积层，参数如下：

层数	层的类型	卷积核/神经元的数量
1	卷积	32 卷积核
2	卷积	32 卷积核
3	卷积	64 卷积核
4	全连接	128 神经元

网络拓扑可以用图 4-14 表示：

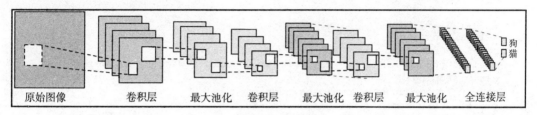

图 4-14　网络拓扑图示

以下代码可以帮助理解参数：

```
# Convolutional Layer 1.
filter_size1 = 3
num_filters1 = 32

# Convolutional Layer 2.
filter_size2 = 3
num_filters2 = 32

# Convolutional Layer 3.
filter_size3 = 3
num_filters3 = 64

# Fully-connected layer.
# Number of neurons in fully-connected layer.
fc_size = 128

# Number of color channels for the images: 1 channel for gray-
scale.
num_channels = 3

# image dimensions (only squares for now)
img_size = 128

# Size of image when flattened to a single dimension
img_size_flat = img_size * img_size * num_channels

# Tuple with height and width of images used to reshape arrays.
img_shape = (img_size, img_size)
```

3）定义类的数量常量（在这个例子中是两个）和其他变量。我们使用斯坦福大学的数据集，为了方便处理，将其减少到狗和猫各 100 个图像。

```
# class info
classes = ['dogs', 'cats']
num_classes = len(classes)
```

```
# batch size
batch_size = 2

# validation split
validation_size = .2
total_iterations = 0
early_stopping = None  # use None if you don't want to implement early stoping
home = '/home/ubuntu/Downloads/dogs_vs_cats'
train_path = home + '/train-cat-dog-100/'
test_path = home + '/test-cat-dog-100/'
checkpoint_dir = home + "/models/"
```

我们首先将数据集读到张量中。读取的逻辑定义在 `dataset` 类中:

```
data = dataset.read_train_sets(train_path, img_size, classes, validation_size=validation_size)
```

这里,我们定义 `train_path`、`image_size`、`classes` 和 `validation_size`。

`read_train_sets(..)` 的实现如下所示:

```
def read_train_sets(train_path, image_size, classes, validation_size=0):
  class DataSets(object):
    pass
  data_sets = DataSets()

  images, labels, ids, cls = load_train(train_path, image_size, classes)
  images, labels, ids, cls = shuffle(images, labels, ids, cls)  # shuffle the data

  if isinstance(validation_size, float):
    validation_size = int(validation_size * images.shape[0])

  validation_images = images[:validation_size]
  validation_labels = labels[:validation_size]
  validation_ids = ids[:validation_size]
  validation_cls = cls[:validation_size]

  train_images = images[validation_size:]
  train_labels = labels[validation_size:]
  train_ids = ids[validation_size:]
  train_cls = cls[validation_size:]

  data_sets.train = DataSet(train_images, train_labels, train_ids, train_cls)
  data_sets.valid = DataSet(validation_images, validation_labels, validation_ids,
```

```
        validation_cls)

    return data_sets
```

这个方法又调用 `load_train(...)` 函数，该函数返回 `numpy.array` 数据类型：

```
def load_train(train_path, image_size, classes):
  images = labels = []
  ids = cls = []
  # load data into arrays
  images = np.array(images)
  labels = np.array(labels)
  ids = np.array(ids)
  cls = np.array(cls)
  return images, labels, ids, cls
```

加载到训练集中的数据是 `validation_set` 的函数；它是从图像数组的第一个维度计算出来的：

> **+** (Ctrl+F1) **images.shape** = {tuple} <type 'tuple'>: (291, 128, 128, 3)

我们用以下代码计算 `validation_size`：

```
validation_size = int(validation_size * images.shape[0])
```

因为 `validation_size` 取的值是 0.2，计算之后为 58.2，取整后的结果为 58：

> **validation_size** = {int} 58

类似地，我们构建测试集——`test_images` 和 `test_ids`：

```
test_images, test_ids = dataset.read_test_set(test_path, img_si
```

这里，`read_test_set(...)` 是一个内部调用的函数：

```
def read_test_set(test_path, image_size):
  images, ids = load_test(test_path, image_size)
  return images, ids
```

`read_trst_set(test_path,image_size)` 轮流调用 `load_tast(test_path, image_size)`，该函数如下所示：

```
def load_test(test_path, image_size):
  path = os.path.join(test_path, '*g')
```

```
    files = sorted(glob.glob(path))

    X_test = []
    X_test_id = []
    print("Reading test images")
    for fl in files:
        flbase = os.path.basename(fl)
        img = cv2.imread(fl)
        img = cv2.resize(img, (image_size, image_size), fx=0.5,
fy=0.5,
            interpolation=cv2.INTER_LINEAR)

        #img = cv2.resize(img, (image_size, image_size),
cv2.INTER_LINEAR)
        X_test.append(img)
        X_test_id.append(flbase)

    ### because we're not creating a DataSet object for the test images,
    ### normalization happens here
    X_test = np.array(X_test, dtype=np.uint8)
    X_test = X_test.astype('float32')
    X_test = X_test / 255
        return X_test, X_test_id
```

4）让我们看一看之前创建的各种 numpy 数组的大小：

```
print("Size of:")
print("- Training-set:\t\t{}".format(len(data.train.labels)))
print("- Test-set:\t\t{}".format(len(test_images)))
print("- Validation-set:\t{}".format(len(data.valid.labels)))

Size of:Size of:
- Training-set:    233
- Test-set:    100
- Validation-set: 58
```

5）在 3×3 的网格中绘制 9 幅随机的图片，每幅图片配有正确的类：

```
images, cls_true = data.train.images, data.train.cls
plot_images(images=images, cls_true=cls_true
```

函数 plot_images 在以下代码块中定义：

```
def plot_images(images, cls_true, cls_pred=None):
    if len(images) == 0:
        print("no images to show")
        return
    else:
        random_indices = random.sample(range(len(images)),
```

```
min(len(images), 9))
    images, cls_true = zip(*[(images[i], cls_true[i]) for i in
random_indices])
    # Create figure with 3x3 sub-plots.
    fig, axes = plt.subplots(3, 3)
    fig.subplots_adjust(hspace=0.3, wspace=0.3)

    for i, ax in enumerate(axes.flat):
        # Plot image.
        print(images[i])
        ax.imshow(images[i].reshape(img_size, img_size,
num_channels))
        print(images[i].size)
        print(img_size)
        print(num_channels)
        # Show true and predicted classes.
        if cls_pred is None:
            xlabel = "True: {0}".format(cls_true[i])
        else:
            xlabel = "True: {0}, Pred: {1}".format(cls_true[i],
cls_pred[i])
        # Show the classes as the label on the x-axis.
        ax.set_xlabel(xlabel)
        # Remove ticks from the plot.
        ax.set_xticks([])
        ax.set_yticks([])
# Ensure the plot is shown correctly with multiple plots
# in a single Notebook cell.
plt.show()
```

代码的输出如图 4-15 所示。

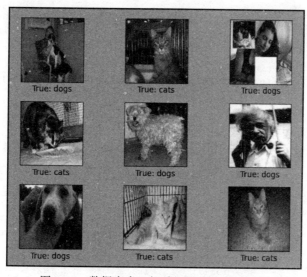

图 4-15　数据库中 9 幅随机图片的测试示例

为输入图片和第一个卷积层定义张量

接下来我们为输入图片和第一个卷积层定义张量。

1. 输入张量

创建一个大小为[None,img_size_flat]的占位符，然后将它的大小变为[-1,img_size,img_size,num_channels]：

```
x = tf.placeholder(tf.float32, shape=[None, img_size_flat], name='x')
x_image = tf.reshape(x, [-1, img_size, img_size, num_channels])
```

参数img_size和num_channels为如下值：

- img_size=128
- num_channels=3

2. 第一个卷积层

在把输入张量的大小变为x_image后，我们构建第一个卷积层。

```
layer_conv1, weights_conv1 = new_conv_layer(input=x_image,
            num_input_channels=num_channels,
                                            filter_size=filter_size1,
                                            num_filters=num_filters1,
                                            use_pooling=True)
print(layer_conv1)
```

在这里定义new_conv_layer(...)函数。我们看看传入这个函数的各个变量的值：

```
num_channels = {int} 3
filter_size1 = {int} 3
num_filters = {int} 32
```

```
def new_conv_layer(input,              # The previous layer.
                   num_input_channels, # Num. channels in prev. layer.
                   filter_size,        # Width and height of each filter.
                   num_filters,        # Number of filters.
                   use_pooling=True):  # Use 2x2 max-pooling.
```

```python
    # Shape of the filter-weights for the convolution.
    # This format is determined by the TensorFlow API.
    shape = [filter_size, filter_size, num_input_channels, num_filters]

    # Create new weights aka. filters with the given shape.
    weights = new_weights(shape=shape)
    # Create new biases, one for each filter.
    biases = new_biases(length=num_filters)

    # Create the TensorFlow operation for convolution.
    # Note the strides are set to 1 in all dimensions.
    # The first and last stride must always be 1,
    # because the first is for the image-number and
    # the last is for the input-channel.
    # But e.g. strides=[1, 2, 2, 1] would mean that the filter
    # is moved 2 pixels across the x- and y-axis of the image.
    # The padding is set to 'SAME' which means the input image
    # is padded with zeroes so the size of the output is the same.
    layer = tf.nn.conv2d(input=input,
                         filter=weights,
                         strides=[1, 1, 1, 1],
                         padding='SAME')

    # Add the biases to the results of the convolution.
    # A bias-value is added to each filter-channel.
    layer += biases

    # Use pooling to down-sample the image resolution?
    if use_pooling:
        # This is 2x2 max-pooling, which means that we
        # consider 2x2 windows and select the largest value
        # in each window. Then we move 2 pixels to the next window.
        layer = tf.nn.max_pool(value=layer,
                               ksize=[1, 2, 2, 1],
                               strides=[1, 2, 2, 1],
                               padding='SAME')

    # Rectified Linear Unit (ReLU).
    # It calculates max(x, 0) for each input pixel x.
    # This adds some non-linearity to the formula and allows us
    # to learn more complicated functions.
    layer = tf.nn.relu(layer)

    # Note that ReLU is normally executed before the pooling,
    # but since relu(max_pool(x)) == max_pool(relu(x)) we can
    # save 75% of the relu-operations by max-pooling first.

    # We return both the resulting layer and the filter-weights
    # because we will plot the weights later.
    return layer, weights
```

运行的时候变量的值如下：

```
+ (Ctrl+F1) biases = {Variable} <tf.Variable 'Variable_1:0' shape=(32,) dtype=float32_ref>

+ (Ctrl+F1) weights = {Variable} <tf.Variable 'Variable:0' shape=(3, 3, 3, 32) dtype=float32_ref>
```

如果运行这段代码，`print(..)` 函数输出如下：

```
Tensor("Relu:0", shape=(?, 64, 64, 32), dtype=float32)
```

输出为第一个卷积层输出张量的大小。

3. 第二个卷积层

在第二个卷积层中，我们将第一个卷积层的输出作为输入，重新构建一个卷积层，参数如下：

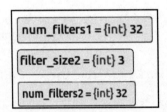

首先，我们为真实的 y 和真实 y 的类别定义占位符：

```
y_true = tf.placeholder(tf.float32, shape=[None, num_classes], name='y_true')

y_true_cls = tf.argmax(y_true, dimension=1)
```

这两个变量的大小如下：

```
+ (Ctrl+F1) y_true = {Tensor} Tensor("y_true:0", shape=(?, 2), dtype=float32)

+ (Ctrl+F1) y_true_cls = {Tensor} Tensor("ArgMax:0", shape=(?,), dtype=int64)
```

```
layer_conv2, weights_conv2 = new_conv_layer(input=layer_conv1,
num_input_channels=num_filters1,filter_size=filter_size2,num_filters=num_fi
```

lters2,use_pooling=True)

以下是它们的值：

❑ num_input_channels＝3
❑ filter_size＝3
❑ num_filters＝32

打印的输出结果如下：

Tensor("Relu_1:0", shape=(?, 32, 32, 32), dtype=float32)

4. 第三个卷积层

该层将第二个卷积层的输出作为输入。我们来看看创建这个层时的输入：

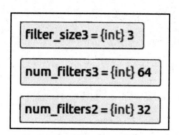

```
shape = [filter_size, filter_size, num_input_channels, num_filters] weights
= new_weights(shape=shape)
```

```
layer_conv3, weights_conv3 = new_conv_layer(input=layer_conv2,
num_input_channels=num_filters2,filter_size=filter_size3,num_filters=num_fi
lters3,use_pooling=True)
```

```
print(layer_conv3)
```

`layer_conv3` 的大小如下：

Tensor("Relu_2:0", shape=(?, 16, 16, 64), dtype=float32)

5. 铺平卷积层

接着，我们把上一层铺平成大小为图像数量乘以特征数量的一层，在本例中为 16 384。如果你注意到了上一层的输出，我们把上一层以如下逻辑进行铺平，16×16×64=16 384：

```
layer_flat, num_features = flatten_layer(layer_conv3)
```

如果打印出这些值，你会看到如下输出：

```
Tensor("Reshape_1:0", shape=(?, 16384), dtype=float32)
16384
```

6. 全连接层

在第四层和第五层，我们定义全连接层：

```
layer_fc1 = new_fc_layer(input=layer_flat,
                         num_inputs=num_features,
                         num_outputs=fc_size,
                         use_relu=True)
```

其中

- `layer_flat`：上一个铺平的层。
- `num_features`：特征数。
- `fc_size`：输出的神经元数量。

下图展示了传给 `new_fc_layer()` 的值：

```
+ (Ctrl+F1) layer_flat = {Tensor} Tensor("Reshape_1:0", shape=(?, 16384), dtype=float32)
num_features = {int} 16384
fc_size = {int} 128
```

```
print(layer_fc1)
```

输出结果如下：

```
Tensor("Relu_3:0", shape=(?, 128), dtype=float32)
```

接着是第二个全连接层，函数参数如下：

- `layer_fc1`：第一个全连接层的输出。
- `num_inputs`：128。

- num_outputs：num_classes，本例中为 2。
- use_relu：指定是否用 relu 激活函数的 bool 值本例中为 False。

```
layer_fc2 = new_fc_layer(input=layer_fc1,
                         num_inputs=fc_size,
                         num_outputs=num_classes,
                         use_relu=False)
```

我们看看第二个全连接层的输出：

```
print(layer_fc2)
Tensor("add_4:0", shape=(?, 2), dtype=float32)
```

7. 定义损失函数和优化器

对 layer_fc2（第二个全连接层）的输出应用 softmax 函数。

在数学中，softmax 函数或归一化的指数函数是逻辑函数的泛化，其将任意的 K 维实数向量 Z 压缩到 K 维实数向量 $\sigma(z)$，$\sigma(z)$ 中的每个值在 [0,1] 之间，并且加起来为 1。

函数由以下公式给出：

$$\sigma(z)j = \frac{e^{z_j}}{\sum_{k=1}^{K} e^{z_k}}, \ j=1,\cdots,K$$

```
y_pred = tf.nn.softmax(layer_fc2)
y_pred_cls = tf.argmax(y_pred, dimension=1)
```

计算交叉熵：

```
cross_entropy = tf.nn.softmax_cross_entropy_with_logits(
  logits=layer_fc2,
  labels=y_true)
cost = tf.reduce_mean(cross_entropy)
```

8. 优化器

接下来我们定义基于 Adam 的优化器。

Adam 和随机梯度下降算法不同。随机梯度下降对于所有的权重更新保持单一的学习速率（称为 α），并且在训练期间学习率不会改变。

Adam 算法对每个网络权重（参数）都会有一个学习率，并且随着训练分开调整学习率。它根据梯度的一阶和二阶矩的估计来计算不同参数各自需要调整的学习速率。

Adam 结合了随机梯度下降法另外两个扩展算法的优点。

自适应梯度算法（Adagrad） 对每个参数都维持一个学习速率，从而改善稀疏梯度的 ML 问题的性能（例如，自然语言和计算机视觉问题）。**均方根传播（RMSProp）** 对每个参数维持一个学习速率；这些学习率是根据最近的权值（参数）的梯度的平均值（梯度变化的速度）来调整的。

```
optimizer = tf.train.AdamOptimizer(learning_rate=1e-4).minimize(cost)
```

我们同样计算 correct_prediction 和 accuracy 的值：

```
correct_prediction = tf.equal(y_pred_cls, y_true_cls)
accuracy = tf.reduce_mean(tf.cast(correct_prediction, tf.float32))
```

9. 第一个迭代周期

初始化会话，调用 optimize() 函数，num_iterations=1：

```
session = tf.Session()
session.run(tf.global_variables_initializer())
batch_size = 2
train_batch_size = batch_size
optimize(num_iterations = 1, data=data, train_batch_size=train_batch_size,
x=x, y_true=y_true,
session=session, optimizer=optimizer, cost=cost, accuracy=accuracy)
```

optimize() 函数在以下代码块中定义：

```
def optimize(num_iterations, data, train_batch_size, x, y_true, session,
optimizer, cost, accuracy):
    # Ensure we update the global variable rather than a local copy.
    global total_iterations

    # Start-time used for printing time-usage below.
    start_time = time.time()
    best_val_loss = float("inf")
    patience = 0

    for i in range(total_iterations,
                   total_iterations + num_iterations):

        # Get a batch of training examples.
```

```
            # x_batch now holds a batch of images and
            # y_true_batch are the true labels for those images.
            x_batch, y_true_batch, _, cls_batch = 
data.train.next_batch(train_batch_size)
            x_valid_batch, y_valid_batch, _, valid_cls_batch = 
data.valid.next_batch(train_batch_size)

            # Convert shape from [num examples, rows, columns, depth]
            # to [num examples, flattened image shape]

            x_batch = x_batch.reshape(train_batch_size, img_size_flat)
            x_valid_batch = x_valid_batch.reshape(train_batch_size, 
img_size_flat)

            # Put the batch into a dict with the proper names
            # for placeholder variables in the TensorFlow graph.
            feed_dict_train = {x: x_batch,
                               y_true: y_true_batch}
            feed_dict_validate = {x: x_valid_batch,
                                  y_true: y_valid_batch}

            # Run the optimizer using this batch of training data.
            # TensorFlow assigns the variables in feed_dict_train
            # to the placeholder variables and then runs the optimizer.
            session.run(optimizer, feed_dict=feed_dict_train)
            # Print status at end of each epoch (defined as full pass through
            # training dataset).
            if i % int(data.train.num_examples/batch_size) == 0:
                val_loss = session.run(cost, feed_dict=feed_dict_validate)
                epoch = int(i / int(data.train.num_examples/batch_size))
                #print_progress(epoch, feed_dict_train, feed_dict_validate, 
val_loss)
                print_progress(session, accuracy, epoch, feed_dict_train, 
feed_dict_validate,
                   val_loss)
                if early_stopping:
                    if val_loss < best_val_loss:
                        best_val_loss = val_loss
                        patience = 0
                    else:
                        patience += 1

                    if patience == early_stopping:
                        break

    # Update the total number of iterations performed.
    total_iterations += num_iterations

    # Ending time.
    end_time = time.time()

    # Difference between start and end-times.
```

```
        time_dif = end_time - start_time

        # Print the time-usage.
        print("Time elapsed: " + str(timedelta(seconds=int(round(time_dif)))))
```

训练集和验证集的准确率以及验证集的损失如下所示：

Epoch 1 --- Training Accuracy: 100.0%, Validation Accuracy: 50.0%, Validation Loss: 0.705

打印 Test-Set 的准确率：

print_validation_accuracy(x, y_true, y_pred_cls, session, data, show_example_errors=True, show_confusion_matrix=False)
Epoch 2 --- Training Accuracy: 50.0%, Validation Accuracy: 100.0%, Validation Loss: 0.320
Accuracy on Test-Set: 43.1% (25 / 58)

接着，迭代 100 次来优化模型：

optimize(num_iterations=100, data=data, train_batch_size=train_batch_size, x=x, y_true=y_true,session=session, optimizer=optimizer, cost=cost, accuracy=accuracy)

print_validation_accuracy(x, y_true, y_pred_cls, session, data, show_example_errors=True,
                          show_confusion_matrix=False)
Accuracy on Test-Set: 62.1% (36 / 58)

从输出中也可以看到猫被错判为狗的例子，如图 4-16 所示。

图 4-16　输出中显示了错判的例子

10. 绘制卷积核以及它们对图片的影响

让我们将两层网络中的卷积核应用到两张测试图片上，看看它们是怎么影响图片的。

```
image1 = test_images[0]
plot_image(image1)
```

`plot_image(image1)`函数的输出如下图所示：

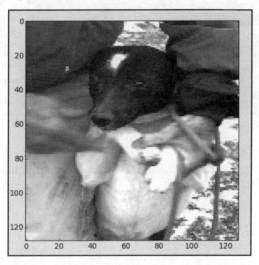

```
image2 = test_images[13]
plot_image(image2)
```

经过卷积核之后的`image2`的输出如下所示：

第一个卷积层： 下图是第一个卷积层权值的图示：

将第一层的卷积核应用到 image1：

```
plot_conv_layer(layer=layer_conv1, image=image1, session=session, x=x)
```

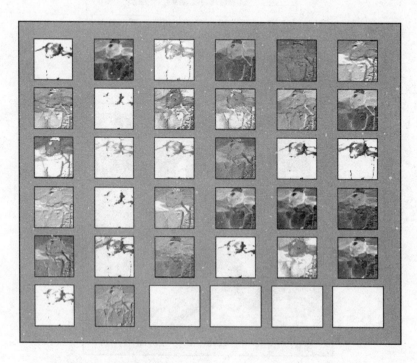

将第一层的卷积核应用到 image2：

```
plot_conv_layer(layer=layer_conv1, image=image2, session=session, x=x)
```

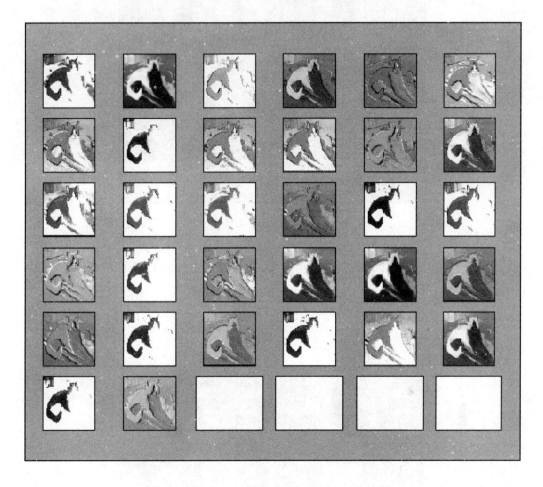

第二个卷积层：现在绘制第二个卷积层的卷积核的权值。第一个卷积层的输出为 16 个通道，这意味着第二个卷积层的输入为 16 个通道。对于每一个输入的通道，第二个卷积层都有一组卷积核权值。我们首先绘制第一个通道的权值。

第二层权值如图 4-17 所示：

```
plot_conv_weights(weights=weights_conv1, session=session)
```

图 4-17 conv2 的权值，输入通道 0，正权值为红色，负权值为蓝色

第二个卷积层有 16 个输入通道，我们可以这样绘制另外的 15 个卷积核权值的图片。如图 4-18 所示，我们仅对第二个通道的权值再绘制一张图。

```
plot_conv_weights(weights=weights_conv2, session=session, input_channel=1)
```

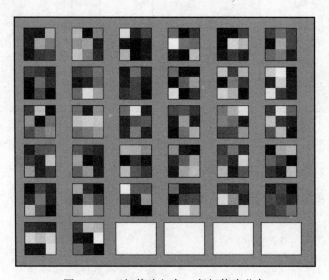

图 4-18 正权值为红色，负权值为蓝色

绘制经过第二个卷积层之后的 image1 和 image2，见图 4-19 和图 4-20：

```
plot_conv_layer(layer=layer_conv2, image=image1, session=session, x=x)
plot_conv_layer(layer=layer_conv2, image=image2, session=session, x=x)
```

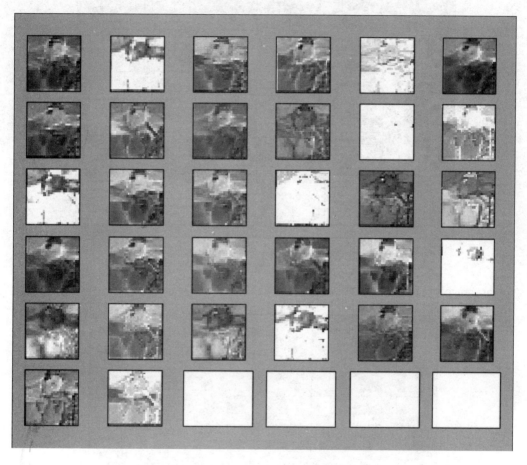

图 4-19　conv2 的权值，输入通道 1。展示了经过第二个卷积层后的 image1

第三个卷积层：让我们绘制第三个卷积层权值的图片；这一层有 64 个卷积核。以下是 iamge1 和 image2 经过这些卷积核时，卷积核权值的图片（见图 4-21 和图 4-22）。

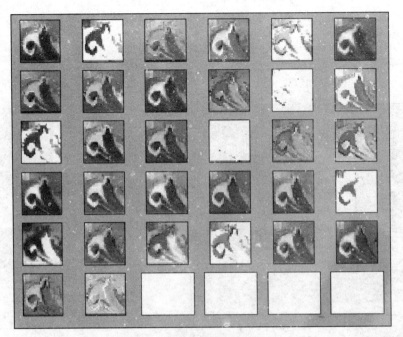

图 4-20 展示了经过第二个卷积层后的 image2

```
plot_conv_weights(weights=weights_conv3, session=session, input_channel=0)
```

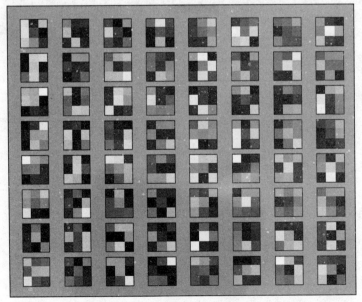

图 4-21 conv2 的权值，输入通道 0。正权值为红色，负权值为蓝色

```
plot_conv_weights(weights=weights_conv3, session=session, input_channel=1)
```

图 4-22 conv2 的权值，输入通道 1。正权值为红色，负权值为蓝色

绘制图片经过第三个卷积层之后的图片：执行以下语句绘制 image1 和 image2 经过第三个卷积层的 64 个卷积核之后的图片，见图 4-23。

```
plot_conv_layer(layer=layer_conv3, image=image1, session=session, x=x)
plot_conv_layer(layer=layer_conv3, image=image2, session=session, x=x)
```

图 4-23 经过 conv3 卷积核后的 image1

图 4-24 是经过第三个卷积层之后的图片。

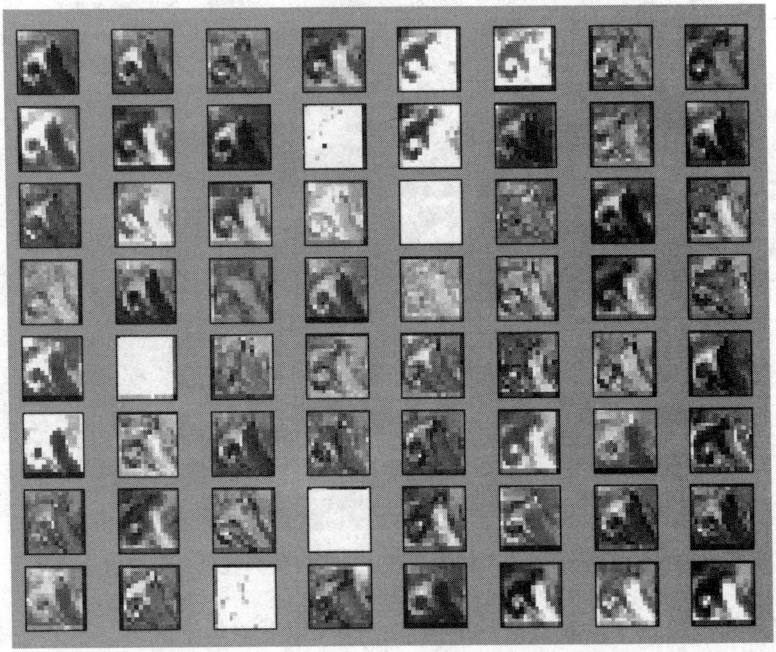

图 4-24　经过 conv3 卷积核之后的 image2

由此，我们完成了对猫与狗数据集的分析，其中我们使用了具有三个隐藏层和两个全连接层的五层 CNN 来构建我们的模型。

4.5　总结

在本章中，你已经学到了卷积的基本操作，而且知道了为什么它在图像预测方面有效。你了解到了诸如步长、填充等概念。接下来是基于斯坦福大学猫与狗的数据集的一个例子。我们使用三个卷积层来构建神经网络，用两个全连接层来展示如何使用它对图像进行分类。我们还绘制了三个卷积层的权重的图，看到了卷积核是如何影响图像的。我们也研究了一些诸如池化的概念，以及它如何使得 CNN 更有效率。

在下一章中，我们将关注一种不同的神经网络，称为**递归神经网络（RNN）**。它用于处理时间序列数据或者在**自然语言处理（NLP）**中预测一个句子的下一个词。

CHAPTER 5

第 5 章

递归神经网络

递归神经网络（RNN）主要用于处理序列或时序数据。在常规的神经网络中，所有的输入层与输出层都是相互独立的。对于一个给定句子去预测下一个单词的任务，我们最好提前知道前面出现过的单词。RNN 对序列中的每个元素重复同样的处理，输出依赖于前面的计算。RNN 可以看作是储存了到目前为止已计算信息的**存储器**。

从前向神经网络到递归神经网络，我们将使用模型不同的组件间权值共享的概念。权值共享将使模型更具有扩展性，并以不同的形式（这里指不同的长度）进行实例应用泛化。

5.1 递归神经网络介绍

为了理解 RNN，我们必须理解前馈神经网络的基本概念。有关前馈神经网络的详细信息请参阅第 3 章。前馈神经网络和递归神经网络都是通过在网络不同节点上执行一系列数学运算来处理信息或特征的。前馈神经网络直接进行信息输入（给定的节点不会输入两次），递归神经网络通过循环来循环输入信息。

对图像数据集训练一个前馈神经网络，目标是将图像预测类别或分类时的损失或误差最小化。利用训练好的超参数或权重集合，神经网络可以对未见过的数据进行分类。

一个训练好的前馈神经网络可以处理任意的图像集合，它对第一个图像的分类并不改变它对其他图像的分类。

简而言之，前馈神经网络没有序列和时序的概念，它们考虑的唯一信息就是当前被要求分类的实例。

RNN 考虑了输入数据的时间特性。RNN 单元的输入既有当前迭代时间的信息也有上一步的信息。详情如图 5-1 所示。

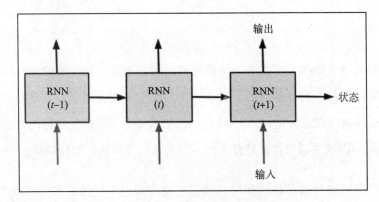

图 5-1　RNN 单元示例

RNN 天然就是重复的，因为它对序列的每个元素做相同的计算，每个输出依赖于上一步计算。用另一种方式来看待 RNN，即 RNN 是有记忆的，它可以捕获截至目前已经计算的信息。RNN 可以利用长序列中的信息和知识，但实际上，它仅能利用前面几步的信息。

一个典型的 RNN 如图 5-2 所示。

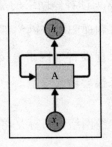

图 5-2　RNN 典型示例

图 5-3 展示了一个展开版本的 RNN，通过展开我们描述一个完整序列的神经网络。想一下有一个五个单词的序列，网络将被展开为一个五层的神经网络，每个单词为一层。

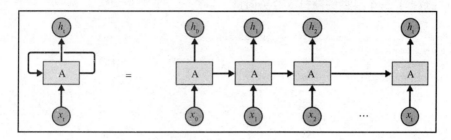

图 5-3 展开的 RNN 示例

RNN 中的计算如图 5-4 所示。

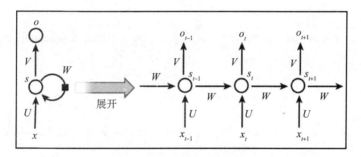

图 5-4 RNN 中的计算图示

- x_t 表示每一步 t 的输入。
- s_t 表示在第 t 步的隐藏状态。隐藏状态相当于网络的记忆。基于上一步的隐藏状态和当前步作为输入计算 s_t，即 $s_t = f(Ux_t + Ws_{t-1})$。
- 函数 f 表示非线性变换，如 `tanh` 或 `ReLU`。第一个隐藏状态通常被初始化为全零。
- o_t 表示第 t 步的输出。在给定句子中预测下一个单词，输出为整个词汇表的概率向量，$o_t = \text{softmax}(Vs_t)$。

5.1.1 RNN 实现

随着程序的运行，会产生一系列数字，目标是基于前面提供的值预测下一个值。RNN 网络每一步的输入是当前值和一个状态向量，状态向量表示神经网络前面几步所看到的信息。该状态向量经 RNN 编码存储，初始设置为零。

训练数据是一个典型的随机二分类向量。

```python
from __future__ import print_function, division
import tensorflow as tf
import numpy as np
import matplotlib.pyplot as plt

"""
define all the constants
"""
numEpochs = 10
seriesLength = 50000
backpropagationLength = 15
stateSize = 4
numClasses = 2
echoStep = 3
batchSize = 5
num_batches = seriesLength // batchSize // backpropagationLength


"""
generate data
"""
def generateData():
    x = np.array(np.random.choice(2, seriesLength, p=[0.5, 0.5]))
    y = np.roll(x, echoStep)
    y[0:echoStep] = 0

    x = x.reshape((batchSize, -1))
    y = y.reshape((batchSize, -1))

    return (x, y)


"""
start computational graph
"""
batchXHolder = tf.placeholder(tf.float32, [batchSize,
backpropagationLength], name="x_input")
batchYHolder = tf.placeholder(tf.int32, [batchSize, backpropagationLength],
name="y_input")

initState = tf.placeholder(tf.float32, [batchSize, stateSize],
"rnn_init_state")

W = tf.Variable(np.random.rand(stateSize+1, stateSize), dtype=tf.float32,
name="weight1")
bias1 = tf.Variable(np.zeros((1,stateSize)), dtype=tf.float32)
```

```python
W2 = tf.Variable(np.random.rand(stateSize, numClasses),dtype=tf.float32,
name="weight2")
bias2 = tf.Variable(np.zeros((1,numClasses)), dtype=tf.float32)

tf.summary.histogram(name="weights", values=W)


# Unpack columns
inputsSeries = tf.unstack(batchXHolder, axis=1, name="input_series")
labelsSeries = tf.unstack(batchYHolder, axis=1, name="labels_series")


# Forward pass
currentState = initState
statesSeries = []
for currentInput in inputsSeries:
    currentInput = tf.reshape(currentInput, [batchSize, 1],
name="current_input")
    inputAndStateConcatenated = tf.concat([currentInput, currentState], 1,
name="input_state_concat")

    nextState = tf.tanh(tf.matmul(inputAndStateConcatenated, W) + bias1,
name="next_state")
    statesSeries.append(nextState)
    currentState = nextState


# calculate loss
logits_series = [tf.matmul(state, W2) + bias2 for state in statesSeries]
predictions_series = [tf.nn.softmax(logits) for logits in logits_series]

losses = [tf.nn.sparse_softmax_cross_entropy_with_logits(labels=labels,
logits=logits) for logits, labels in zip(logits_series,labelsSeries)]
total_loss = tf.reduce_mean(losses, name="total_loss")

train_step = tf.train.AdagradOptimizer(0.3).minimize(total_loss,
name="training")


"""
plot computation
"""
def plot(loss_list, predictions_series, batchX, batchY):
    plt.subplot(2, 3, 1)
    plt.cla()
    plt.plot(loss_list)

    for batchSeriesIdx in range(5):
        oneHotOutputSeries = np.array(predictions_series)[:,
batchSeriesIdx, :]
        singleOutputSeries = np.array([(1 if out[0] < 0.5 else 0) for out
```

```
           in oneHotOutputSeries])
                plt.subplot(2, 3, batchSeriesIdx + 2)
                plt.cla()
                plt.axis([0, backpropagationLength, 0, 2])
                left_offset = range(backpropagationLength)
                plt.bar(left_offset, batchX[batchSeriesIdx, :], width=1,
color="blue")
                plt.bar(left_offset, batchY[batchSeriesIdx, :] * 0.5, width=1,
color="red")
                plt.bar(left_offset, singleOutputSeries * 0.3, width=1,
color="green")

        plt.draw()
        plt.pause(0.0001)


    """
    run the graph
    """
    with tf.Session() as sess:
        writer = tf.summary.FileWriter("logs", graph=tf.get_default_graph())
        sess.run(tf.global_variables_initializer())
        plt.ion()
        plt.figure()
        plt.show()
        loss_list = []

        for epoch_idx in range(numEpochs):
            x,y = generateData()
            _current_state = np.zeros((batchSize, stateSize))

            print("New data, epoch", epoch_idx)

            for batch_idx in range(num_batches):
                start_idx = batch_idx * backpropagationLength
                end_idx = start_idx + backpropagationLength

                batchX = x[:,start_idx:end_idx]
                batchY = y[:,start_idx:end_idx]

                _total_loss, _train_step, _current_state, _predictions_series =
sess.run(
                    [total_loss, train_step, currentState, predictions_series],
                    feed_dict={
                        batchXHolder:batchX,
                        batchYHolder:batchY,
                        initState:_current_state
                    })

                loss_list.append(_total_loss)

                # fix the cost summary later
```

```
            tf.summary.scalar(name="totalloss", tensor=_total_loss)

            if batch_idx%100 == 0:
                print("Step",batch_idx, "Loss", _total_loss)
                plot(loss_list, _predictions_series, batchX, batchY)

plt.ioff()
plt.show()
```

计算图

计算图如图 5-5 所示。

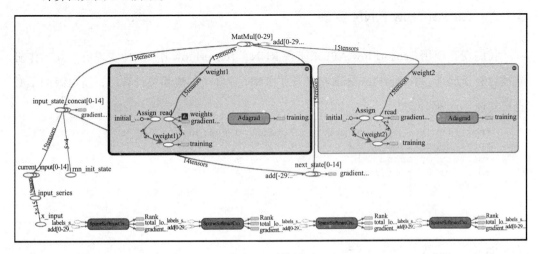

图 5-5　计算图示

输出结果如下：

```
New data, epoch 0
Step 0 Loss 0.777418
Step 600 Loss 0.693907
New data, epoch 1
Step 0 Loss 0.690996
Step 600 Loss 0.691115
New data, epoch 2
Step 0 Loss 0.69259
Step 600 Loss 0.685826
New data, epoch 3
Step 0 Loss 0.684189
Step 600 Loss 0.690608
New data, epoch 4
Step 0 Loss 0.691302
Step 600 Loss 0.691309
New data, epoch 5
Step 0 Loss 0.69172
```

```
Step 600 Loss 0.694034
New data, epoch 6
Step 0 Loss 0.692927
Step 600 Loss 0.42796
New data, epoch 7
Step 0 Loss 0.42423
Step 600 Loss 0.00845207
New data, epoch 8
Step 0 Loss 0.188478
Step 500 Loss 0.00427217
```

5.1.2 TensorFlow RNN 实现

我们现在使用 TensorFlow API 来实现 RNN，TensorFlow 中 RNN 实现的内部工作机制被封装了起来。TensorFlow rnn 包实现了 RNN 并自动创建相关计算图，所以我们不必自己实现 for 循环。

```
from __future__ import print_function, division
import tensorflow as tf
import numpy as np
import matplotlib.pyplot as plt

"""
define all the constants
"""
numEpochs = 10
seriesLength = 50000
backpropagationLength = 15
stateSize = 4
numClasses = 2
echoStep = 3
batchSize = 5
num_batches = seriesLength // batchSize // backpropagationLength
"""
generate data
"""
def generateData():
    x = np.array(np.random.choice(2, seriesLength, p=[0.5, 0.5]))
    y = np.roll(x, echoStep)
    y[0:echoStep] = 0

    x = x.reshape((batchSize, -1))
    y = y.reshape((batchSize, -1))

    return (x, y)
"""
start computational graph
"""
batchXHolder = tf.placeholder(tf.float32, [batchSize,
```

```python
                        backpropagationLength], name="x_input")
batchYHolder = tf.placeholder(tf.int32, [batchSize, backpropagationLength],
name="y_input")

initState = tf.placeholder(tf.float32, [batchSize, stateSize],
"rnn_init_state")

W = tf.Variable(np.random.rand(stateSize+1, stateSize), dtype=tf.float32,
name="weight1")
bias1 = tf.Variable(np.zeros((1,stateSize)), dtype=tf.float32)

W2 = tf.Variable(np.random.rand(stateSize, numClasses),dtype=tf.float32,
name="weight2")
bias2 = tf.Variable(np.zeros((1,numClasses)), dtype=tf.float32)

tf.summary.histogram(name="weights", values=W)


# Unpack columns
inputsSeries = tf.split(axis=1, num_or_size_splits=backpropagationLength,
value=batchXHolder)
labelsSeries = tf.unstack(batchYHolder, axis=1)

# Forward passes
from tensorflow.contrib import rnn
cell = rnn.BasicRNNCell(stateSize)
statesSeries, currentState = rnn.static_rnn(cell, inputsSeries, initState)

# calculate loss
logits_series = [tf.matmul(state, W2) + bias2 for state in statesSeries]
predictions_series = [tf.nn.softmax(logits) for logits in logits_series]

losses = [tf.nn.sparse_softmax_cross_entropy_with_logits(labels=labels,
logits=logits) for logits, labels in zip(logits_series,labelsSeries)]
total_loss = tf.reduce_mean(losses, name="total_loss")

train_step = tf.train.AdagradOptimizer(0.3).minimize(total_loss,
name="training")


"""
plot computation
"""
def plot(loss_list, predictions_series, batchX, batchY):
    plt.subplot(2, 3, 1)
    plt.cla()
    plt.plot(loss_list)

    for batchSeriesIdx in range(5):
        oneHotOutputSeries = np.array(predictions_series)[:,
batchSeriesIdx, :]
        singleOutputSeries = np.array([(1 if out[0] < 0.5 else 0) for out
in oneHotOutputSeries])

        plt.subplot(2, 3, batchSeriesIdx + 2)
        plt.cla()
```

```python
        plt.axis([0, backpropagationLength, 0, 2])
        left_offset = range(backpropagationLength)
        plt.bar(left_offset, batchX[batchSeriesIdx, :], width=1, color="blue")
        plt.bar(left_offset, batchY[batchSeriesIdx, :] * 0.5, width=1, color="red")
        plt.bar(left_offset, singleOutputSeries * 0.3, width=1, color="green")

    plt.draw()
    plt.pause(0.0001)


"""
run the graph
"""
with tf.Session() as sess:
    writer = tf.summary.FileWriter("logs", graph=tf.get_default_graph())
    sess.run(tf.global_variables_initializer())
    plt.ion()
    plt.figure()
    plt.show()
    loss_list = []

    for epoch_idx in range(numEpochs):
        x,y = generateData()
        _current_state = np.zeros((batchSize, stateSize))

        print("New data, epoch", epoch_idx)

        for batch_idx in range(num_batches):
            start_idx = batch_idx * backpropagationLength
            end_idx = start_idx + backpropagationLength

            batchX = x[:,start_idx:end_idx]
            batchY = y[:,start_idx:end_idx]

            _total_loss, _train_step, _current_state, _predictions_series = sess.run(
                [total_loss, train_step, currentState, predictions_series],
                feed_dict={
                    batchXHolder:batchX,
                    batchYHolder:batchY,
                    initState:_current_state
                })

            loss_list.append(_total_loss)

            # fix the cost summary later
            tf.summary.scalar(name="totalloss", tensor=_total_loss)

            if batch_idx%100 == 0:
                print("Step",batch_idx, "Loss", _total_loss)
                plot(loss_list, _predictions_series, batchX, batchY)

plt.ioff()
plt.show()
```

计算图

计算图如图 5-6 所示。

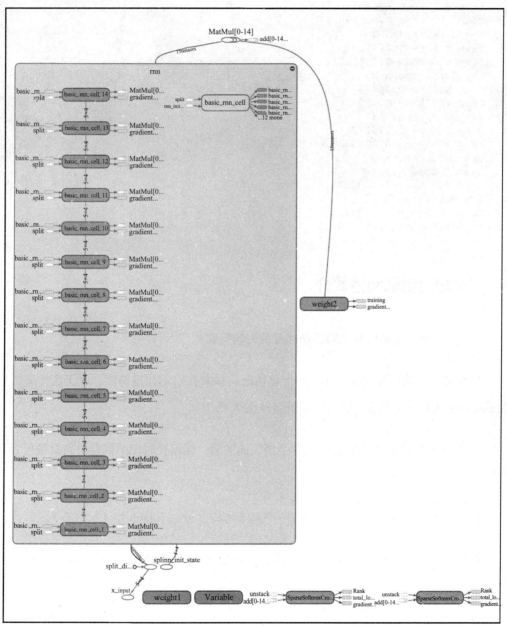

图 5-6　TensorFlow RNN 计算图示

输出结果如下:

```
New data, epoch 0
Step 0 Loss 0.688437
Step 600 Loss 0.00107078
New data, epoch 1
Step 0 Loss 0.214923
Step 600 Loss 0.00111716
New data, epoch 2
Step 0 Loss 0.214962
Step 600 Loss 0.000730697
New data, epoch 3
Step 0 Loss 0.276177
Step 600 Loss 0.000362316
New data, epoch 4
Step 0 Loss 0.1641
Step 600 Loss 0.00025342
New data, epoch 5
Step 0 Loss 0.0947087
Step 600 Loss 0.000276762
```

5.2 长短期记忆网络简介

递归神经网络面临的最大障碍是梯度消失的问题。

梯度展示了所有权重随着误差的变化情况,如果我们不知道梯度,就无法按降低损失或误差的方向更新权重,神经网络就会停止学习。

长短期记忆(LSTM)旨在克服梯度消失的问题。保留长时间的信息对其隐藏行为是极其有效的。

在标准的 RNN 中,循环单元会有一个基本结构,如一个 tanh 层(见图 5-7)。

如图 5-7 所示,LSTM 也有一个类似链状的结构,但是循环单元结构不同(见图 5-8)。

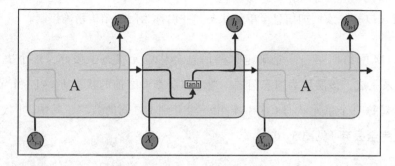

图 5-7　RNN 中的 tanh 层图示

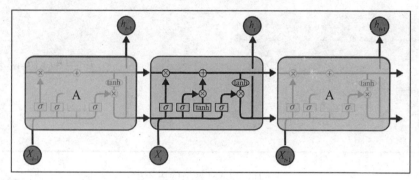

图 5-8　含链式结构的 LSTM 图示

5.2.1　LSTM 的生命周期

LSTM 的核心是单元状态,它就像一个传送带。传送带在微小的线性相互作用下移动。数据流动非常直接,见图 5-9。

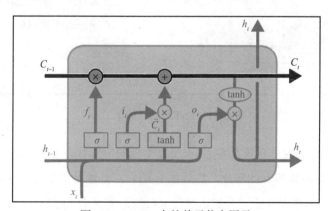

图 5-9　LSTM 中的单元状态图示

LSTM 可以移除或增加信息至单元状态,并由称为门的结构精确控制。

1) LSTM 网络的第一步是确定哪些信息会从单元状态中丢弃,这个决策由一个 sigmoid 层来完成,该层称为遗忘门层。遗忘门层查看之前的状态 $h(t-1)$ 和当前的输入 $x(t)$,并针对每一个单元状态 $C(t-1)$ 输出一个 0 到 1 之间的数字,其中,1 代表**完全保留**,0 代表**完全丢弃**(见图 5-10)。

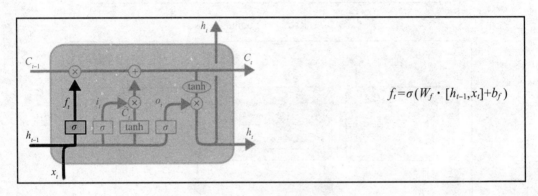

图 5-10　LSTM 网络的第一步

2) 下一步是确定单元状态需要保留的新信息(见图 5-11)。首先,sigmoid 层的输入门决定哪些值将被更新。其次,tanh 层生成一个可以添加到状态的新候选值 C 的向量。

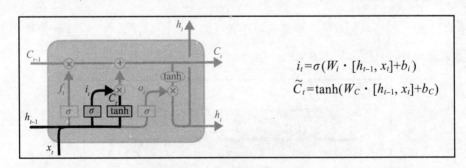

图 5-11　LSTM 网络的第二步

3) 我们现在将旧的单元状态 $C(t-1)$ 更新为新的单元状态 $C(t)$。将旧状态乘以 $f(t)$,并且遗忘我们之前决定遗忘的部分。然后加上 $i(t)*\tilde{C}$,这是对的新候选值进行缩放的结果,缩放的程度表示我们决定对状态值进行更新的力度大小(见图 5-12)。

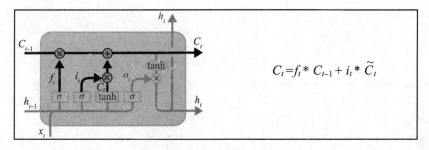

图 5-12　LSTM 网络的第三步

4）最后，我们决定输出，这是对状态值进行过滤或修改的值。首先，经过 sigmoid 层来确定我们要输出状态的哪些部分。接下来，通过 tanh 将单元状态推到 −1 和 1 之间，并将其乘以 sigmoid 门的输出，这样只输出我们确定想要的部分（见图 5-13）。

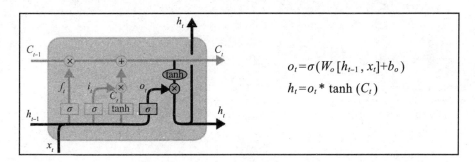

图 5-13　LSTM 网络的最后一步

5.2.2　LSTM 实现

LSTM 记忆、遗忘并选择要传递的信息，然后根据当前状态和输入进行输出。一个 LSTM 由多个组件构成，但使用 TensorFlow API，可以很容易地实现 LSTM：

```
from __future__ import print_function, division
import tensorflow as tf
import numpy as np
import matplotlib.pyplot as plt
from tensorflow.contrib import rnn

"""
define all the constants
"""
numEpochs = 10
seriesLength = 50000
```

```python
backpropagationLength = 15
stateSize = 4
numClasses = 2
echoStep = 3
batchSize = 5
num_batches = seriesLength // batchSize // backpropagationLength


"""
generate data
"""
def generateData():
    x = np.array(np.random.choice(2, seriesLength, p=[0.5, 0.5]))
    y = np.roll(x, echoStep)
    y[0:echoStep] = 0
    x = x.reshape((batchSize, -1))
    y = y.reshape((batchSize, -1))

    return (x, y)


"""
start computational graph
"""
batchXHolder = tf.placeholder(tf.float32, [batchSize,
backpropagationLength], name="x_input")
batchYHolder = tf.placeholder(tf.int32, [batchSize, backpropagationLength],
name="y_input")

# rnn replace
#initState = tf.placeholder(tf.float32, [batchSize, stateSize],
"rnn_init_state")

cellState = tf.placeholder(tf.float32, [batchSize, stateSize])
hiddenState = tf.placeholder(tf.float32, [batchSize, stateSize])
initState = rnn.LSTMStateTuple(cellState, hiddenState)

W = tf.Variable(np.random.rand(stateSize+1, stateSize), dtype=tf.float32,
name="weight1")
bias1 = tf.Variable(np.zeros((1,stateSize)), dtype=tf.float32)

W2 = tf.Variable(np.random.rand(stateSize, numClasses),dtype=tf.float32,
name="weight2")
bias2 = tf.Variable(np.zeros((1,numClasses)), dtype=tf.float32)

tf.summary.histogram(name="weights", values=W)


# Unpack columns
inputsSeries = tf.split(axis=1, num_or_size_splits=backpropagationLength,
value=batchXHolder)
labelsSeries = tf.unstack(batchYHolder, axis=1)
```

```python
# Forward passes

# rnn replace
# cell = rnn.BasicRNNCell(stateSize)
# statesSeries, currentState = rnn.static_rnn(cell, inputsSeries, initState)

cell = rnn.BasicLSTMCell(stateSize, state_is_tuple=True)
statesSeries, currentState = rnn.static_rnn(cell, inputsSeries, initState)
# calculate loss
logits_series = [tf.matmul(state, W2) + bias2 for state in statesSeries]
predictions_series = [tf.nn.softmax(logits) for logits in logits_series]

losses = [tf.nn.sparse_softmax_cross_entropy_with_logits(labels=labels,
logits=logits) for logits, labels in zip(logits_series,labelsSeries)]
total_loss = tf.reduce_mean(losses, name="total_loss")

train_step = tf.train.AdagradOptimizer(0.3).minimize(total_loss,
name="training")


"""
plot computation
"""
def plot(loss_list, predictions_series, batchX, batchY):
    plt.subplot(2, 3, 1)
    plt.cla()
    plt.plot(loss_list)

    for batchSeriesIdx in range(5):
        oneHotOutputSeries = np.array(predictions_series)[:, batchSeriesIdx, :]
        singleOutputSeries = np.array([(1 if out[0] < 0.5 else 0) for out in oneHotOutputSeries])

        plt.subplot(2, 3, batchSeriesIdx + 2)
        plt.cla()
        plt.axis([0, backpropagationLength, 0, 2])
        left_offset = range(backpropagationLength)
        plt.bar(left_offset, batchX[batchSeriesIdx, :], width=1, color="blue")
        plt.bar(left_offset, batchY[batchSeriesIdx, :] * 0.5, width=1, color="red")
        plt.bar(left_offset, singleOutputSeries * 0.3, width=1, color="green")

    plt.draw()
    plt.pause(0.0001)


"""
```

```python
run the graph
"""
with tf.Session() as sess:
    writer = tf.summary.FileWriter("logs", graph=tf.get_default_graph())
    sess.run(tf.global_variables_initializer())
    plt.ion()
    plt.figure()
    plt.show()
    loss_list = []

    for epoch_idx in range(numEpochs):
        x,y = generateData()

        # rnn remove
        # _current_state = np.zeros((batchSize, stateSize))

        _current_cell_state = np.zeros((batchSize, stateSize))
        _current_hidden_state = np.zeros((batchSize, stateSize))

        print("New data, epoch", epoch_idx)

        for batch_idx in range(num_batches):
            start_idx = batch_idx * backpropagationLength
            end_idx = start_idx + backpropagationLength

            batchX = x[:,start_idx:end_idx]
            batchY = y[:,start_idx:end_idx]

            _total_loss, _train_step, _current_state, _predictions_series = sess.run(
                [total_loss, train_step, currentState, predictions_series],
                feed_dict={
                    batchXHolder:batchX,
                    batchYHolder:batchY,
                    cellState: _current_cell_state,
                    hiddenState: _current_hidden_state
                })

            _current_cell_state, _current_hidden_state = _current_state

            loss_list.append(_total_loss)

            # fix the cost summary later
            tf.summary.scalar(name="totalloss", tensor=_total_loss)

            if batch_idx%100 == 0:
                print("Step",batch_idx, "Loss", _total_loss)
                plot(loss_list, _predictions_series, batchX, batchY)

    plt.ioff()
    plt.show()
```

计算图

图 5-14 来自 TensorBoard，它展示了 LSTM 是如何工作的。

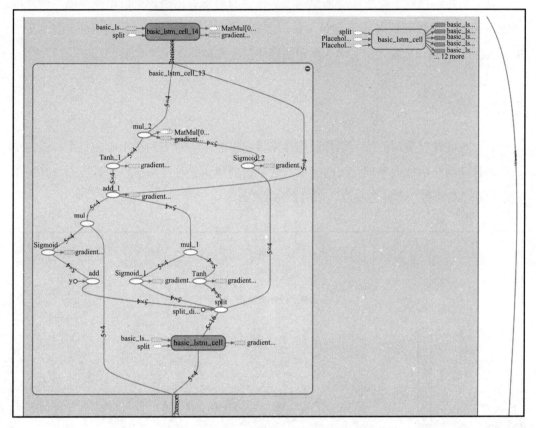

图 5-14　来自 TensorBoard 的计算图

输出如下所示：

```
New data, epoch 0
Step 0 Loss 0.696803
Step 600 Loss 0.00743465
New data, epoch 1
Step 0 Loss 0.404039
Step 600 Loss 0.00243205
New data, epoch 2
Step 0 Loss 1.11536
Step 600 Loss 0.00140995
New data, epoch 3
Step 0 Loss 0.858743
Step 600 Loss 0.00141037
```

5.3 情感分析

我们现在编写一个应用程序来预测电影评论的情感。用户评论由单词序列组成。单词顺序的编码对预测情绪很有价值。第一步是做词嵌入（word embedding）。第二步是 RNN 以向量序列作为输入，并考虑向量的顺序以生成预测。

5.3.1 词嵌入

我们现在针对单词训练一个神经网络来做向量表示。对于句子中心的每个特定的单词，即神经网络的输入单词，我们观察与之临近的单词，神经网络将告诉我们所选择单词附近在字典中每个单词的概率（见图 5-15）。

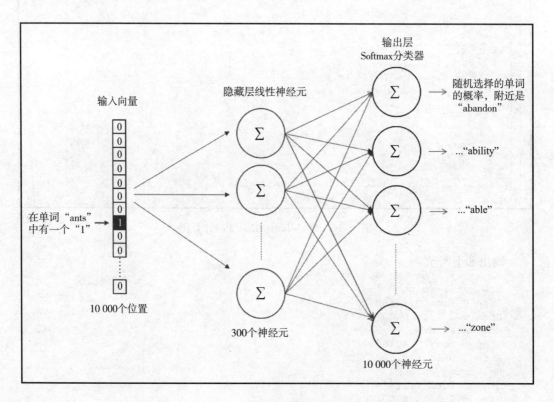

图 5-15 临近单词的概率计算图示

```python
import time
import tensorflow as tf
import numpy as np
import utility
from tqdm import tqdm
from urllib.request import urlretrieve
from os.path import isfile, isdir
import zipfile
from collections import Counter
import random

dataDir = 'data'
dataFile = 'text8.zip'
datasetName = 'text 8 data set'

'''
track progress of file download
'''


class DownloadProgress(tqdm):
    lastBlock = 0
    def hook(self, blockNum=1, blockSize=1, totalSize=None):
        self.total = totalSize
        self.update((blockNum - self.lastBlock) * blockSize)
        self.lastBlock = blockNum


if not isfile(dataFile):
    with DownloadProgress(unit='B', unit_scale=True, miniters=1, desc=datasetName) as progressBar:
        urlretrieve('http://mattmahoney.net/dc/text8.zip', dataFile, progressBar.hook)

if not isdir(dataDir):
    with zipfile.ZipFile(dataFile) as zipRef:
        zipRef.extractall(dataDir)

with open('data/text8') as f:
    text = f.read()

'''
pre process the downloaded wiki text
'''
words = utility.preProcess(text)
print(words[:30])

print('Total words: {}'.format(len(words)))
print('Unique words: {}'.format(len(set(words))))

'''
```

```
convert words to integers
'''
int2vocab, vocab2int = utility.lookupTable(words)
intWords = [vocab2int[word] for word in words]
print('test')

'''
sub sampling (***think of words as int's***)
'''
threshold = 1e-5
wordCounts = Counter(intWords)
totalCount = len(intWords)
frequency = {word: count / totalCount for word, count in
wordCounts.items()}
probOfWords = {word: 1 - np.sqrt(threshold / frequency[word]) for word in
wordCounts}
trainWords = [word for word in intWords if random.random() < (1 -
probOfWords[word])]
'''
get window batches
'''


def getTarget(words, index, windowSize=5):
    rNum = np.random.randint(1, windowSize + 1)
    start = index - rNum if (index - rNum) > 0 else 0
    stop = index + rNum
    targetWords = set(words[start:index] + words[index + 1:stop + 1])

    return list(targetWords)


'''
Create a generator of word batches as a tuple (inputs, targets)
'''


def getBatches(words, batchSize, windowSize=5):
    nBatches = len(words) // batchSize
    print('no. of batches {}'.format(nBatches))

    # only full batches
    words = words[:nBatches * batchSize]

    start = 0
    for index in range(0, len(words), batchSize):
        x = []
        y = []
        stop = start + batchSize
        batchWords = words[start:stop]
        for idx in range(0, len(batchWords), 1):
```

```
            yBatch = getTarget(batchWords, idx, windowSize)
            y.extend(yBatch)
            x.extend([batchWords[idx]] * len(yBatch))
        start = stop + 1
        yield x, y

'''
start computational graph
'''
train_graph = tf.Graph()
with train_graph.as_default():
    netInputs = tf.placeholder(tf.int32, [None], name='inputS')
    netLabels = tf.placeholder(tf.int32, [None, None], name='labelS')
'''
create embedding layer
'''
nVocab = len(int2vocab)
nEmbedding = 300
with train_graph.as_default():
    embedding = tf.Variable(tf.random_uniform((nVocab, nEmbedding), -1, 1))
    embed = tf.nn.embedding_lookup(embedding, netInputs)


'''
Below, create weights and biases for the softmax layer. Then, use
tf.nn.sampled_softmax_loss to calculate the loss
'''
n_sampled = 100
with train_graph.as_default():
    soft_W = tf.Variable(tf.truncated_normal((nVocab, nEmbedding)))
    soft_b = tf.Variable(tf.zeros(nVocab), name="softmax_bias")

    # Calculate the loss using negative sampling
    loss = tf.nn.sampled_softmax_loss(
        weights=soft_W,
        biases=soft_b,
        labels=netLabels,
        inputs=embed,
        num_sampled=n_sampled,
        num_classes=nVocab)

    cost = tf.reduce_mean(loss)
    optimizer = tf.train.AdamOptimizer().minimize(cost)

'''
Here we're going to choose a few common words and few uncommon words. Then,
we'll print out the closest words to them.
It's a nice way to check that our embedding table is grouping together
words with similar semantic meanings.
'''
with train_graph.as_default():
```

```python
        validSize = 16
        validWindow = 100

        validExamples = np.array(random.sample(range(validWindow), validSize //
2))
        validExamples = np.append(validExamples,
                                  random.sample(range(1000, 1000 +
validWindow), validSize // 2))

        validDataset = tf.constant(validExamples, dtype=tf.int32)
        norm = tf.sqrt(tf.reduce_sum(tf.square(embedding), 1, keep_dims=True
        normalizedEmbedding = embedding / norm
        valid_embedding = tf.nn.embedding_lookup(normalizedEmbedding,
validDataset)
        similarity = tf.matmul(valid_embedding,
tf.transpose(normalizedEmbedding))


'''
Train the network. Every 100 batches it reports the training loss. Every
1000 batches, it'll print out the validation
words.
'''
epochs = 10
batch_size = 1000
window_size = 10

with train_graph.as_default():
    saver = tf.train.Saver()

with tf.Session(graph=train_graph) as sess:
    iteration = 1
    loss = 0
    sess.run(tf.global_variables_initializer())

    for e in range(1, epochs + 1):
        batches = getBatches(trainWords, batch_size, window_size)
        start = time.time()
        for x, y in batches:

            feed = {netInputs: x,
                    netLabels: np.array(y)[:, None]}
            trainLoss, _ = sess.run([cost, optimizer], feed_dict=feed)

            loss += trainLoss

            if iteration % 100 == 0:
                end = time.time()
                print("Epoch {}/{}".format(e, epochs),
                      "Iteration: {}".format(iteration),
                      "Avg. Training loss: {:.4f}".format(loss / 100),
                      "{:.4f} sec/batch".format((end - start) / 100))
```

```
            loss = 0
            start = time.time()

        if iteration % 1000 == 0:
            sim = similarity.eval()
            for i in range(validSize):
                validWord = int2vocab[validExamples[i]]
                topK = 8
                nearest = (-sim[i, :]).argsort()[1:topK + 1]
                log = 'Nearest to %s:' % validWord
                for k in range(topK):
                    closeWord = int2vocab[nearest[k]]
                    logStatement = '%s %s,' % (log, closeWord)
                print(logStatement)

        iteration += 1
    save_path = saver.save(sess, "checkpoints/text8.ckpt")
    embed_mat = sess.run(normalizedEmbedding)


'''
Restore the trained network if you need to
'''
with train_graph.as_default():
    saver = tf.train.Saver()

with tf.Session(graph=train_graph) as sess:
    saver.restore(sess, tf.train.latest_checkpoint('checkpoints'))
    embed_mat = sess.run(embedding)


'''
Below we'll use T-SNE to visualize how our high-dimensional word vectors
cluster together. T-SNE is used to project
these vectors into two dimensions while preserving local structure.
'''
import matplotlib.pyplot as plt
from sklearn.manifold import TSNE
vizWords = 500
tsne = TSNE()
embedTSNE = tsne.fit_transform(embed_mat[:vizWords, :])

fig, ax = plt.subplots(figsize=(14, 14))
for idx in range(vizWords):
    plt.scatter(*embedTSNE[idx, :], color='steelblue')
    plt.annotate(int2vocab[idx], (embedTSNE[idx, 0], embedTSNE[idx, 1]), alpha=0.7)
```

输出结果如下：

```
Total words: 16680599
 Unique words: 63641
 no. of batches 4626
Epoch 1/10 Iteration: 100 Avg. Training loss: 21.7284 0.3363 sec/batch
 Epoch 1/10 Iteration: 1000 Avg. Training loss: 20.2269 0.3668 sec/batch

Nearest to but: universities, hungry, kyu, grandiose, edema, patty, stores,
psychometrics,
 Nearest to three: sulla, monuc, conjuring, ontological, auf, grimoire,
unpredictably, frenetic,

Nearest to world: turkle, spectroscopic, jules, servicio, sportswriter,
kamikazes, act, earns,
Epoch 1/10 Iteration: 1100 Avg. Training loss: 20.1983 0.3650 sec/batch
 Epoch 1/10 Iteration: 2000 Avg. Training loss: 19.1581 0.3767 sec/batch

Nearest to but: universities, hungry, edema, kyu, grandiose, stores, patty,
psychometrics,
 Nearest to three: monuc, sulla, unpredictably, grimoire, hickey,
ontological, conjuring, rays,
 Nearest to world: turkle, spectroscopic, jules, sportswriter, kamikazes,
alfons, servicio, act,
 ......
```

5.3.2　使用 RNN 进行情感分析

以下示例展示了如何使用 RNN 进行情感分析，电影评论数据编码为固定长度的整数值。然后转换为词嵌入（嵌入向量），以循环的方式输入到 LSTM 层中，最终选择最后的预测结果作为情感输出结果。

```
import numpy as np
import tensorflow as tf
from string import punctuation
from collections import Counter

'''
movie review dataset for sentiment analysis
'''
with open('data/reviews.txt', 'r') as f:
    movieReviews = f.read()
with open('data/labels.txt', 'r') as f:
    labels = f.read()


'''
data cleansing - remove punctuations
'''
text = ''.join([c for c in movieReviews if c not in punctuation])
```

```
movieReviews = text.split('\n')

text = ' '.join(movieReviews)
words = text.split()

print(text[:500])
print(words[:100])


'''
build a dictionary that maps words to integers
'''
counts = Counter(words)
vocabulary = sorted(counts, key=counts.get, reverse=True)
vocab2int = {word: i for i, word in enumerate(vocabulary, 1)}

reviewsInts = []
for review in movieReviews:
    reviewsInts.append([vocab2int[word] for word in review.split()])


'''
convert labels from positive and negative to 1 and 0 respectively
'''
labels = labels.split('\n')
labels = np.array([1 if label == 'positive' else 0 for label in labels])

reviewLengths = Counter([len(x) for x in reviewsInts])
print("Min review length are: {}".format(reviewLengths[0]))
print("Maximum review length are: {}".format(max(reviewLengths)))


'''
remove the review with zero length from the reviewsInts list
'''
nonZeroIndex = [i for i, review in enumerate(reviewsInts) if len(review) != 0]
print(len(nonZeroIndex))


'''
turns out its the final review that has zero length. But that might not
always be the case, so let's make it more
general.
'''
reviewsInts = [reviewsInts[i] for i in nonZeroIndex]
labels = np.array([labels[i] for i in nonZeroIndex])
'''
create an array features that contains the data we'll pass to the network.
The data should come from reviewInts, since
we want to feed integers to the network. Each row should be 200 elements
long. For reviews shorter than 200 words,
left pad with 0s. That is, if the review is ['best', 'movie', 'renaira'],
```

```python
[100, 40, 20] as integers, the row will look
like [0, 0, 0, ..., 0, 100, 40, 20]. For reviews longer than 200, use on
the first 200 words as the feature vector.
'''
seqLen = 200
features = np.zeros((len(reviewsInts), seqLen), dtype=int)
for i, row in enumerate(reviewsInts):
    features[i, -len(row):] = np.array(row)[:seqLen]

print(features[:10,:100])


'''
lets create training, validation and test data sets. trainX and trainY for example.
also define a split percentage function 'splitPerc' as the percentage of data to keep in the training
set. usually this is 0.8 or 0.9.
'''
splitPrec = 0.8
splitIndex = int(len(features)*0.8)
trainX, valX = features[:splitIndex], features[splitIndex:]
trainY, valY = labels[:splitIndex], labels[splitIndex:]

testIndex = int(len(valX)*0.5)
valX, testX = valX[:testIndex], valX[testIndex:]
valY, testY = valY[:testIndex], valY[testIndex:]

print("Train set: {}".format(trainX.shape), "\nValidation set: {}".format(valX.shape), "\nTest set: {}".format(testX.shape))
print("label set: {}".format(trainY.shape), "\nValidation label set: {}".format(valY.shape), "\nTest label set: {}".format(testY.shape))


'''
tensor-flow computational graph
'''
lstmSize = 256
lstmLayers = 1
batchSize = 500
learningRate = 0.001
nWords = len(vocab2int) + 1

# create graph object and add nodes to the graph
graph = tf.Graph()

with graph.as_default():
    inputData = tf.placeholder(tf.int32, [None, None], name='inputData')
    labels = tf.placeholder(tf.int32, [None, None], name='labels')
    keepProb = tf.placeholder(tf.float32, name='keepProb')


'''
```

```
let us create the embedding layer (word2vec)
'''
# number of neurons in hidden or embedding layer
embedSize = 300

with graph.as_default():
    embedding = tf.Variable(tf.random_uniform((nWords, embedSize), -1, 1))
    embed = tf.nn.embedding_lookup(embedding, inputData)


'''
lets use tf.contrib.rnn.BasicLSTMCell to create an LSTM cell, later add drop out to it with
tf.contrib.rnn.DropoutWrapper. and finally create multiple LSTM layers with
tf.contrib.rnn.MultiRNNCell.
'''
with graph.as_default():
    with tf.name_scope("RNNLayers"):
        def createLSTMCell():
            lstm = tf.contrib.rnn.BasicLSTMCell(lstmSize, reuse=tf.get_variable_scope().reuse)
            return tf.contrib.rnn.DropoutWrapper(lstm, output_keep_prob=keepProb)

        cell = tf.contrib.rnn.MultiRNNCell([createLSTMCell() for _ in range(lstmLayers)])

        initialState = cell.zero_state(batchSize, tf.float32)


'''
set tf.nn.dynamic_rnn to add the forward pass through the RNN. here we're actually passing in vectors from the
embedding layer 'embed'.
'''
with graph.as_default():
    outputs, finalState = tf.nn.dynamic_rnn(cell, embed, initial_state=initialState)


'''
final output will carry the sentiment prediction, therefore lets get the last output with outputs[:, -1],
the we calculate the cost from that and labels.
'''
with graph.as_default():
    predictions = tf.contrib.layers.fully_connected(outputs[:, -1], 1, activation_fn=tf.sigmoid)
    cost = tf.losses.mean_squared_error(labels, predictions)

    optimizer = tf.train.AdamOptimizer(learningRate).minimize(cost)


'''
```

```
now we can add a few nodes to calculate the accuracy which we'll use in the
validation pass.
'''
with graph.as_default():
    correctPred = tf.equal(tf.cast(tf.round(predictions), tf.int32),
labels)
    accuracy = tf.reduce_mean(tf.cast(correctPred, tf.float32))

'''
get batches
'''
def getBatches(x, y, batchSize=100):
    nBatches = len(x) // batchSize
    x, y = x[:nBatches * batchSize], y[:nBatches * batchSize]
    for i in range(0, len(x), batchSize):
        yield x[i:i + batchSize], y[i:i + batchSize]

'''
training phase
'''
epochs = 1

with graph.as_default():
    saver = tf.train.Saver()

with tf.Session(graph=graph) as sess:
    writer = tf.summary.FileWriter("logs", graph=tf.get_default_graph())
    sess.run(tf.global_variables_initializer())
    iteration = 1
    for e in range(epochs):
        state = sess.run(initialState)

        for i, (x, y) in enumerate(getBatches(trainX, trainY, batchSize),
1):
            feed = {inputData: x, labels: y[:, None], keepProb: 0.5,
initialState: state}

            loss, state, _ = sess.run([cost, finalState, optimizer],
feed_dict=feed)

            if iteration % 5 == 0:
                print("Epoch are: {}/{}".format(e, epochs), "Iteration is:
{}".format(iteration), "Train loss is: {:.3f}".format(loss))

            if iteration % 25 == 0:
                valAcc = []
                valState = sess.run(cell.zero_state(batchSize, tf.float32))
                for x, y in getBatches(valX, valY, batchSize):
                    feed = {inputData: x, labels: y[:, None], keepProb: 1,
initialState: valState}
                    batchAcc, valState = sess.run([accuracy, finalState],
feed_dict=feed)
                    valAcc.append(batchAcc)
                print("Val acc: {:.3f}".format(np.mean(valAcc)))
            iteration += 1
```

```
            saver.save(sess, "checkpoints/sentimentanalysis.ckpt")
        saver.save(sess, "checkpoints/sentimentanalysis.ckpt")

'''
testing phase
'''
testAcc = []
with tf.Session(graph=graph) as sess:
    saver.restore(sess, "checkpoints/sentiment.ckpt")

    testState = sess.run(cell.zero_state(batchSize, tf.float32))
    for i, (x, y) in enumerate(getBatches(testY, testY, batchSize), 1):
        feed = {inputData: x,
                labels: y[:, None],
                keepProb: 1,
                initialState: testState}
        batchAcc, testState = sess.run([accuracy, finalState], 
feed_dict=feed)
        testAcc.append(batchAcc)
    print("Test accuracy is: {:.3f}".format(np.mean(testAcc)))
```

计算图

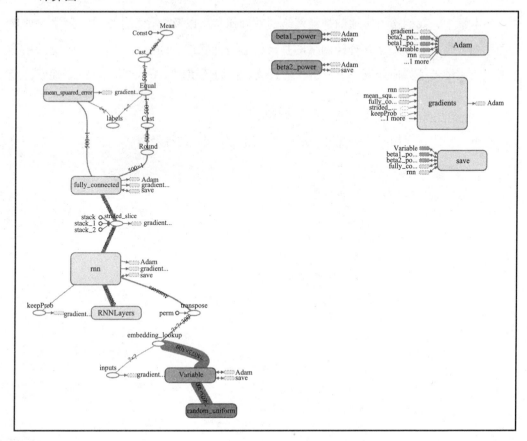

输出结果如下:

```
Train set: (20000, 200)
 Validation set: (2500, 200)
 Test set: (2500, 200)
 label set: (20000,)
 Validation label set: (2500,)
 Test label set: (2500,)
Val acc: 0.682
 Val acc: 0.692
 Val acc: 0.714
 Val acc: 0.808
 Val acc: 0.763
 Val acc: 0.826
 Val acc: 0.854
 Val acc: 0.872
```

5.4 总结

在本章中,我们学习了递归神经网络的基础概念,以及为什么递归神经网络对处理时间序列数据很有效。递归神经网络的基础概念包括状态、词嵌入、长期记忆等。接下来通过实例学习了如何开发一个情感分析系统。我们还使用 TensorFlow 实现了递归神经网络。

在下一章中,我们将看一看另一种称为"生成模型"的神经网络。

CHAPTER 6

第 6 章

生成模型

生成模型是机器学习模型的一个分类,它们用来描述数据如何生成。为了训练生成模型,首先在任意域中累积大量数据,然后训练模型来构建或生成类似数据。

换句话说,这些模型可以学习构建类似我们给它们的数据。一种方法是使用**生成对抗网络(GAN)**,这将作为本章的一部分进行详细讨论。

本章将介绍以下主题:

- 生成模型的简介。
- 生成对抗网络。

6.1 生成模型简介

有两种机器学习模型:生成模型和判别模型。我们看看下面的分类器列表:决策树、神经网络、随机森林、广义增强模型、逻辑回归、朴素贝叶斯和**支持向量机(SVM)**。其中大部分是分类器和组合模型。这里比较特殊的是朴素贝叶斯。这是列表中唯一的生成模型。其他都是判别模型的例子。

生成模型和判别模型的根本区别在于潜在概率的推导形式。在本章中,我们将研究一些重要概念,比如生成模型类型和 GAN,但在此之前,先来了解生成模型和判别模型之间的一些关键区别。

6.1.1 判别模型对生成模型

判别模型学习目标变量 Y 和特征 X 的条件概率 $P(Y|X)$。这是最小二乘回归的工作方式，它是我们使用的一种推理模式。这是理清变量之间的关系的一种方法。

生成模型旨在对数据集进行完整的概率描述。生成模型的目标是计算联合概率 $P(X, Y)$，可以直接计算或通过计算 $P(Y|X)$ 和 $P(X)$ 来计算联合概率分布 $P(X, Y)$，然后推断需要分类的新数据的条件概率。这种方法需要比回归更扎实的概率信息，但它提供了数据概率结构的完整模型。了解联合概率分布可以生成数据；因此，朴素贝叶斯是一个生成模型。

假设我们有一个有监督的学习任务，其中 x_i 是数据给定的特征，y_i 是相应的标签。一种基于特征 x 预测 y 的方法是从 (x_i, y_i) 中学习一个函数 $f()$，它的输入是 x，输出是最有可能的 y。这些模型属于判别模型的范畴，因为你学习的是如何区分不同类别的对应的 x。支持向量机和神经网络等方法属于这一类。即使你能够非常准确地对数据进行分类，你也不知道数据是如何生成的。

第二种方法是模拟如何生成数据并学习函数 $f(x, y)$，该函数给出一个分数，这个分数是由 x 和 y 一起决定的。然后，对于新的 x，通过找到使 $f(x, y)$ 最大的 y 就可以确定 x 的标签 y。一个典型的例子是高斯混合模型。

另一个例子是：你可以把 x 想象成一幅图像，而 y 在图像中是一种像狗一样的物体。概率 $p(y|x)$ 告诉我们，给定一幅输入图像和它可能属于的所有类别，模型认为这幅图是狗的概率有多大。试图直接对这个概率进行建模的算法称为**判别模型**。

另一方面，生成模型试图学习被称为联合概率 $p(y, x)$ 的函数。我们可以这样理解这个模型，它代表了 x 是一个图像，同时图像中有一只狗 y 的概率是多少。这两个概率是相关的，可以写成 $p(y, x) = p(x)p(y|x)$，其中 $p(x)$ 是输入 x 是图像的概率。概率 $p(x)$ 在

文献中通常被称为**密度函数**。

称这些模型为生成模型的主要原因是它们可以同时获得输入和输出的概率。利用这一点，我们可以通过对动物种类 y 和新图像 x 进行抽样，然后根据概率 $p(y, x)$ 生成动物图像。

我们可以主要学习仅依赖于输入空间的密度函数 $p(x)$。

两种模型都很有用。然而，相对来说，生成模型与判别模型相比有一个有趣的优势，那就是即使没有标签，它们也有可能理解和解释输入数据的潜藏结构。

这在实际生活中是非常有用的。

6.1.2 生成模型的类型

在机器学习领域最近取得的成功案例中，判别模型处在前列。尽管模型不能生成新的样本或数据，但模型可以对给定的输入进行预测。

生成模型最新进展背后的思想是将生成问题转换为预测问题，然后使用深度学习算法来学习这样一个问题。

1. 自编码器

将生成问题转换为判别问题的一种方法是学习输入空间到自身的映射。例如，我们想要学习一个恒等映射，对于每个图像 x，在理想情况下，其预测值是它本身，即 $x=f(x)$，其中 f 是预测模型。

在现在这个形式下，模型可能没有什么用，但是我们可以从中构建一个生成模型。

在这里，我们构建一个由两个主要组件组成的模型：一个编码器模型 $q(h \mid x)$，将输入映射到另一个空间（称为隐藏空间或者潜在空间，用 h 表示）；一个解码器模型 $q(x \mid h)$，从隐藏的输入空间中学习相反的映射。

这些组件（编码器和解码器）连接在一起形成一个端到端的可训练模型。编码器和解码器模型都是结构不同的神经网络，例如 RNN 和注意力网络，以获得期望的结果。

模型学习完之后，我们可以从编码器中移除解码器，然后单独使用它们。为了生成新的数据样本，我们可以首先从潜藏空间生成数据，然后将其送到解码器中，最后从输出空间产生新样本。自编码器在第 8 章中有更详细的介绍。

2. GAN

正如在上一节看到的那样，我们可以想一个通用概念来构建在某种关系下可以一起工作的神经网络，并且通过训练它们以此帮我们学习到使我们可以生成新的数据样本的潜藏空间。

另一种类型的生成网络是 GAN，我们有一个生成器模型 $q(x \mid h)$ 将低维潜藏空间 h（通常表示为简单分布的噪声）映射到输入空间 x。这与解码器在自编码器中的作用非常相似。

现在的目标是要引入一个判别模型 $p(y \mid x)$，它可以将一个输入实例 x 与一个是/否的二元回答 y 相关联，这个回答是关于 x 是生成器模型根据输入生成的数据或者是我们训练数据集中的真实样本。

让我们以之前的图像为例。假设生成器模型创建一个新的图像，我们也有实际数据集中的真实图像。如果生成器模型是正确的，判别器模型将不能容易地区分这两个图像。如果生成器模型很差，那么判别器模型很容易分辨哪一个是假的或者伪造的，哪一个是真的。

当这两种模型相互耦合时，我们可以以端到端的形式训练它们。假设随着时间的推移，生成器模型可以更好地欺骗判别器模型，同时，我们在更困难的分辨虚假的问题上训练判别器模型。最后，我们有了一个生成器模型，判别器很难将它的输出和用于训练的实际数据进行区分。

刚开始训练时，判别器模型可以很容易地分辨出真实数据集中的样本和由生成器生成的样本。由于生成器模型生成的数据越来越好，我们开始看到越来越多的看起来与数据集相似的生成样本。以下示例描述了随时间学习的 GAN 模型的生成图像（见图 6-1）。

图 6-1　随着时间学习的 GAN 模型的生成图像

在接下来的小节中，我们将详细讨论各种 GAN。

3. 序列模型

如果数据本质上是有时序的，那么我们可以使用称为**序列模型**的特定算法。这些模

型可以学习形如 $p(y|x\_n, x\_1)$ 的概率，其中 i 是表示序列中位置的参数，x_i 是第 i 个输入样本。

举个例子，我们可以把每个单词当作是一系列的字符组成，每个句子当作是一系列的单词组成，而每一个段落都是一系列的句子组成。输出 y 可以是句子的情感。

使用与自编码器类似的技巧，我们可以将 y 替换为系列或序列中的下一个元素，即 $y=x\_n+1$，从而让模型进行学习。

6.2 GAN

GAN 是由 Ian Goodfellow 领导的蒙特利尔大学的一组研究人员提出的。GAN 模型的核心思想是建立两种互相竞争的神经网络模型。第一种网络模型将噪声作为输入并产生样本（因此称为**生成器**）。第二种模型（称为**判别器**）从生成器和实际训练数据中获取样本，并且应该能够区分这两种来源。生成网络和判别网络在玩一个连续的游戏，其中生成器模型正在学习生成更真实的样本或示例，而判别器模型正在学习如何更好地区分生成的数据和真实数据。两种网络同时训练，目标是互相竞争，使生成的样本与真实数据不可区分（见图 6-2）。

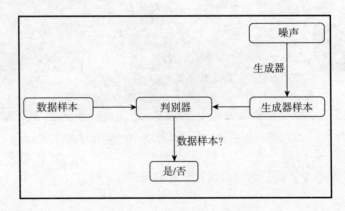

图 6-2　两种互相竞争的网络模型

用这样的类比描述 GAN，生成器模型就像一个试图产生某种伪造材料的伪造者，判别器模型是警察，试图检测伪造的物品。这看起来有点类似于强化学习，其中生成器通过判别器获得奖励，从而使其知道生成的数据是否准确。GAN 的关键区别在于，我们可以将梯度信息从判别器网络反向传播回生成器网络，使得生成器知道如何调整其参数以生成可以欺骗判别器的输出数据。

到目前为止，GAN 主要应用于生成自然图像。它们在图像生成任务中的结果是最好的，并且比其他基于最大似然训练方法生成的图像更清晰。

图 6-3 是一些由 GAN 生成的图像示例。

图 6-3　由 GAN 生成的图像示例

6.2.1　GAN 示例

为了更深入地理解 GAN 的工作原理，我们在 TensorFlow 中使用 GAN 解决一个简单的问题，即估计一维高斯分布。

首先，我们将创建实际的数据分布——一个简单的高斯分布，均值为 4，标准差为 0.5。它有一个采样函数，用于从分布中返回给定数量的样本（按值排序）。我们学习的数据分布如图 6-4 所示。

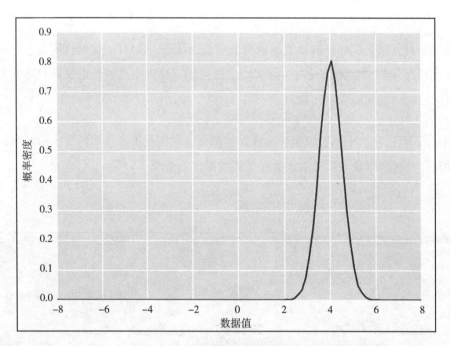

图 6-4　一维高斯分布的数据分布图示

生成器的输入噪声的分布也用一个类似的用于实际数据的采样函数来定义。

生成器和判别器网络都非常简单。生成器首先进行线性变换，然后经过一个非线性变换（softplus 函数），接着进行另一个线性变换。

我们保持判别器比生成器强。否则，它就没有足够的能力来学习如何准确区分生成样本和真实样本。

因此，我们让判别器的网络深度更深，维度更高。除了最后一层是 sigmoid 函数（其输出可以被描述为一个概率），我们在所有层中使用了 tanh 非线性函数。

我们将这些网络连接在一起，作为 TensorFlow 图的一部分，并为每个网络定义损失函数，这样生成器网络只要欺骗判别器网络就可以了。使用 TensorFlow 中学习率指数衰减的梯度下降优化方法作为优化方法。

为了训练模型，我们从数据分布和噪声分布中分别抽取样本，并且交替优化生成器和判别器的参数。

我们将看到，在训练开始时，生成器产生了与真实数据非常不同的分布。网络在收敛到一个集中在输入分布平均值的较窄分布之前，慢慢地学着和真实分布变得非常相近。训练完网络之后，这两种分布如图 6-5 所示。

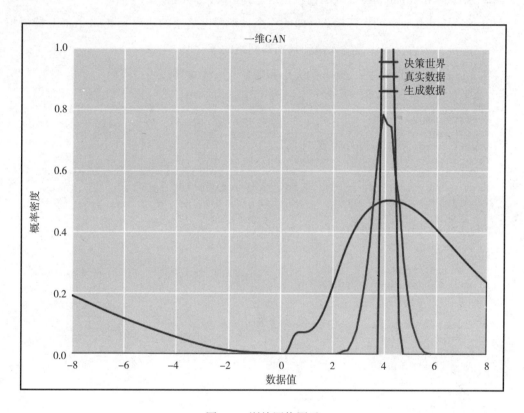

图 6-5　训练网络图示

生成网络无法收敛到合理的参数，即它会生成一个很窄的分布或者一些点的模样，这个是 GAN 训练失败的主要原因。解决方法是允许判别器一次查看多个样本，这种技术称为 minibatch 判别。minibatch 判别是一种可以查看整个 batch 样本然后确定它们是来自生成器网络还是来自实际数据的方法。

该方法总结如下：

- 取出判别器网络任意中间层的输出。
- 用三维张量乘以这个输出生成一个大小为 numOfKernels×kernelDim 的矩阵。
- 计算一个 batch 中所有样本矩阵中行之间的 L1 距离，然后再对它们应用负指数函数。
- 样本的 minibatch 特征或属性就是这些指数化距离的总和。
- 将 minibatch 层的实际输入（即前一个判别器网络层的输出）和 minibatch 特征相连接，然后将其作为输入送到判别器网络的下一层。

在 minibatch 判别器中，batch 的大小和超参数一样重要。

```python
import argparse
import numpy as np
import tensorflow as tf
import matplotlib.pyplot as plt
from matplotlib import animation
import seaborn as sns
from tensorflow.python.training.gradient_descent import GradientDescentOptimizer

sns.set(color_codes=True)

seed = 42
np.random.seed(seed)
tf.set_random_seed(seed)


# gaussian data distribution
class DataDist(object):
    def __init__(self):
        self.mue = 4
        self.sigma = 0.5

    def sample(self, N):
        samples = np.random.normal(self.mue, self.sigma, N)
        samples.sort()
        return samples


# data distribution with noise
class GeneratorDist(object):
    def __init__(self, rnge):
        self.rnge = rnge

    def sample(self, N):
        return np.linspace(-self.rnge, self.rnge, N) + \
            np.random.random(N) * 0.01
```

```python
# linear method
def linearUnit(input, output_dim, scope=None, stddev=1.0):
    with tf.variable_scope(scope or 'linearUnit'):
        weight = tf.get_variable(
            'weight',
            [input.get_shape()[1], output_dim],
            initializer=tf.random_normal_initializer(stddev=stddev)
        )
        bias = tf.get_variable(
            'bias',
            [output_dim],
            initializer=tf.constant_initializer(0.0)
        )
        return tf.matmul(input, weight) + bias


# generator network
def generatorNetwork(input, hidden_size):
    hidd0 = tf.nn.softplus(linearUnit(input, hidden_size, 'g0'))
    hidd1 = linearUnit(hidd0, 1, 'g1')
    return hidd1


# discriminator network
def discriminatorNetwork(input, h_dim, minibatch_layer=True):
    hidd0 = tf.nn.relu(linearUnit(input, h_dim * 2, 'd0'))
    hidd1 = tf.nn.relu(linearUnit(hidd0, h_dim * 2, 'd1'))

    if minibatch_layer:
        hidd2 = miniBatch(hidd1)
    else:
        hidd2 = tf.nn.relu(linearUnit(hidd1, h_dim * 2, scope='d2'))

    hidd3 = tf.sigmoid(linearUnit(hidd2, 1, scope='d3'))
    return hidd3


# minibatch
def miniBatch(input, numKernels=5, kernelDim=3):
    x = linearUnit(input, numKernels * kernelDim, scope='minibatch', stddev=0.02)
    act = tf.reshape(x, (-1, numKernels, kernelDim))
    differences = tf.expand_dims(act, 3) - \
            tf.expand_dims(tf.transpose(act, [1, 2, 0]), 0)
    absDiffs = tf.reduce_sum(tf.abs(differences), 2)
    minibatchFeatures = tf.reduce_sum(tf.exp(-absDiffs), 2)
    return tf.concat([input, minibatchFeatures], 1)


# optimizer
def optimizer(loss, var_list):
    learning_rate = 0.001
```

```python
        step = tf.Variable(0, trainable=False)
        optimizer = tf.train.AdamOptimizer(learning_rate).minimize(
            loss,
            global_step=step,
            var_list=var_list
        )
        return optimizer


# log
def log(x):
    return tf.log(tf.maximum(x, 1e-5))


class GAN(object):
    def __init__(self, params):
        with tf.variable_scope('Generator'):
            self.zee = tf.placeholder(tf.float32, shape=(params.batchSize, 1))
            self.Gee = generatorNetwork(self.zee, params.hidden_size)

        self.xVal = tf.placeholder(tf.float32, shape=(params.batchSize, 1))
        with tf.variable_scope('Discriminator'):
            self.Dis1 = discriminatorNetwork(
                self.xVal,
                params.hidden_size,
                params.minibatch
            )
        with tf.variable_scope('D', reuse=True):
            self.Dis2 = discriminatorNetwork(
                self.Gee,
                params.hidden_size,
                params.minibatch
            )

        self.lossD = tf.reduce_mean(-log(self.Dis1) - log(1 - self.Dis2))
        self.lossG = tf.reduce_mean(-log(self.Dis2))

        vars = tf.trainable_variables()
        self.dParams = [v for v in vars if v.name.startswith('D/')]
        self.gParams = [v for v in vars if v.name.startswith('G/')]

        self.optD = optimizer(self.lossD, self.dParams)
        self.optG = optimizer(self.lossG, self.gParams)


'''
Train GAN model
'''
def trainGan(model, data, gen, params):
    animFrames = []

    with tf.Session() as session:
```

```
            tf.local_variables_initializer().run()
            tf.global_variables_initializer().run()

            for step in range(params.numSteps + 1):
                x = data.sample(params.batchSize)
                z = gen.sample(params.batchSize)
                lossD, _, = session.run([model.lossD, model.optD], {
                    model.x: np.reshape(x, (params.batchSize, 1)),
                    model.z: np.reshape(z, (params.batchSize, 1))
                })

                z = gen.sample(params.batchSize)
                lossG, _ = session.run([model.lossG, model.optG], {
                    model.z: np.reshape(z, (params.batchSize, 1))
                })

                if step % params.log_every == 0:
                    print('{}: {:.4f}\t{:.4f}'.format(step, lossD, lossG))

                if params.animPath and (step % params.animEvery == 0):
                    animFrames.append(
                        getSamples(model, session, data, gen.range,
params.batchSize)
                    )

        if params.animPath:
            saveAnimation(animFrames, params.animPath, gen.range)
        else:
            samps = getSamples(model, session, data, gen.range,
params.batchSize)
            plotDistributions(samps, gen.range)
def getSamples(
        model,
        session,
        data,
        sampleRange,
        batchSize,
        numPoints=10000,
        numBins=100
):
    xs = np.linspace(-sampleRange, sampleRange, numPoints)
    binss = np.linspace(-sampleRange, sampleRange, numBins)

    # decision boundary
    db = np.zeros((numPoints, 1))
    for i in range(numPoints // batchSize):
        db[batchSize * i:batchSize * (i + 1)] = session.run(
            model.D1,
            {
                model.x: np.reshape(
                    xs[batchSize * i:batchSize * (i + 1)],
```

```
                            (batchSize, 1)
                        )
                    }
                )

        # data distribution
        d = data.sample(numPoints)
        pds, _ = np.histogram(d, bins=binss, density=True)

        zs = np.linspace(-sampleRange, sampleRange, numPoints)
        g = np.zeros((numPoints, 1))
        for i in range(numPoints // batchSize):
            g[batchSize * i:batchSize * (i + 1)] = session.run(
                model.G,
                {
                    model.z: np.reshape(
                        zs[batchSize * i:batchSize * (i + 1)],
                        (batchSize, 1)
                    )
                }
            )
        pgs, _ = np.histogram(g, bins=binss, density=True)

        return db, pds, pgs


def plotDistributions(samps, sampleRange):
    db, pd, pg = samps
    dbX = np.linspace(-sampleRange, sampleRange, len(db))
    pX = np.linspace(-sampleRange, sampleRange, len(pd))
    f, ax = plt.subplots(1)
    ax.plot(dbX, db, label='Decision Boundary')
    ax.set_ylim(0, 1)
    plt.plot(pX, pd, label='Real Data')
    plt.plot(pX, pg, label='Generated Data')
    plt.title('1D Generative Adversarial Network')
    plt.xlabel('Data Values')
    plt.ylabel('Probability Density')
    plt.legend()
    plt.show()


def saveAnimation(animFrames, animPath, sampleRange):
    f, ax = plt.subplots(figsize=(6, 4))
    f.suptitle('1D GAN', fontsize=15)
    plt.xlabel('dataValues')
    plt.ylabel('probabilityDensity')
    ax.set_xlim(-6, 6)
    ax.set_ylim(0, 1.4)
    lineDb, = ax.plot([], [], label='decision boundary')
    linePd, = ax.plot([], [], label='real data')
```

```python
        linePg, = ax.plot([], [], label='generated data')
        frameNumber = ax.text(
            0.02,
            0.95,
            '',
            horizontalalignment='left',
            verticalalignment='top',
            transform=ax.transAxes
        )
        ax.legend()

        db, pd, _ = animFrames[0]
        dbX = np.linspace(-sampleRange, sampleRange, len(db))
        pX = np.linspace(-sampleRange, sampleRange, len(pd))

        def init():
            lineDb.set_data([], [])
            linePd.set_data([], [])
            linePg.set_data([], [])
            frameNumber.set_text('')
            return (lineDb, linePd, linePg, frameNumber)

        def animate(i):
            frameNumber.set_text(
                'Frame: {}/{}'.format(i, len(animFrames))
            )
            db, pd, pg = animFrames[i]
            lineDb.set_data(dbX, db)
            linePd.set_data(pX, pd)
            linePg.set_data(pX, pg)
            return (lineDb, linePd, linePg, frameNumber)

        anim = animation.FuncAnimation(
            f,
            animate,
            init_func=init,
            frames=len(animFrames),
            blit=True
        )
        anim.save(animPath, fps=30, extra_args=['-vcodec', 'libx264'])


# start gan modeling
def gan(args):
    model = GAN(args)
    trainGan(model, DataDist(), GeneratorDist(range=8), args)


# input arguments
def parseArguments():
    argParser = argparse.ArgumentParser()
```

```python
    argParser.add_argument('--num-steps', type=int, default=5000,
                           help='the number of training steps to take')
    argParser.add_argument('--hidden-size', type=int, default=4,
                           help='MLP hidden size')
    argParser.add_argument('--batch-size', type=int, default=8,
                           help='the batch size')
    argParser.add_argument('--minibatch', action='store_true',
                           help='use minibatch discrimination')
    argParser.add_argument('--log-every', type=int, default=10,
                           help='print loss after this many steps')
    argParser.add_argument('--anim-path', type=str, default=None,
                           help='path to the output animation file')
    argParser.add_argument('--anim-every', type=int, default=1,
                           help='save every Nth frame for animation')
    return argParser.parse_args()


# start the gan app
if __name__ == '__main__':
    gan(parseArguments())
```

输出结果：

```
 0: 6.6300 0.1449
 10: 6.5390 0.1655
 20: 6.4552 0.1866
 30: 6.3789 0.2106
 40: 6.3190 0.2372
 50: 6.2814 0.2645
 60: 6.2614 0.2884
 70: 6.2556 0.3036
 80: 6.2814 0.3104
 90: 6.2796 0.3113
100: 6.3008 0.3106
110: 6.2923 0.3112
120: 6.2792 0.3153
130: 6.3299 0.3196
140: 6.3512 0.3205
150: 6.2999 0.3197
160: 6.3513 0.3236
170: 6.3521 0.3291
180: 6.3377 0.3292
```

6.2.2　GAN 的种类

接下来的一小节介绍几种不同的 GAN，例如 Vanilla GAN、Conditional GAN 等。详情请见 https://arxiv.org。

以下关于各 GAN 网络的描述取自 https://arxiv.org 网站上相应的论文。

1. Vanilla GAN

Vanilla GAN 有两个网络，分别称为生成器网络和判别器网络。

两个网络同时进行训练，并在最小化最大值的过程中互相竞争和对抗。生成器网络是已经经过训练或预先准备的，因此它可以通过由输入生成的图像来欺骗判别器网络，判别器网络训练的目标是不被生成器网络欺骗。

有关 Vanilla GAN 的进一步阅读请参阅 https://arxiv.org/abs/1406.2661。

2. Conditional GAN

GAN 最初作为一种新型生成模型训练的方法。Conditional GAN 是使用额外标签数据的 GAN 网络。它生成的图像质量十分优秀，并能够在一定程度上控制生成的图像的外观。这个模型可以用来学习多模式模型。

有关 Conditional GAN 的进一步阅读请参阅 https://arxiv.org/abs/1411.1784。

3. Info GAN

Info GAN 能够以无监督的方式编码或学习重要的图像特征或分离式表征。例如，对数字的旋转进行编码。InfoGAN 最大化潜变量的一个子集与观测值之间的互信息。

有关 Info GAN 的进一步阅读请参阅 https://arxiv.org/abs/1606.03657。

4. Wasserstein GAN

WGAN 是传统 GAN 训练的一个可选方案。WGAN 的损失函数与图像质量有关。此外，训练的稳定性得到了提升，而且稳定性不依赖于网络结构，并提供学习曲线，这对调试网络很有用。

有关Wasserstein GAN的进一步阅读请参阅 https://arxiv.org/abs/1701.07875。

5. Coupled GAN

Coupled GAN用于在两个独立的域中生成一组相似的图像。它由一组GAN组成，每个GAN负责在单个域中生成图像。Coupled GAN从两个域中的图像学习一个联合分布，这些图像是从每个单独的域的边缘分布中分别采样出来的。

 有关Coupled GAN的进一步阅读请参阅 https://arxiv.org/abs/1606.07536。

6.3 总结

生成模型是一个快速发展的学习研究领域。随着我们持续优化模型并增加训练时间和数据集，我们可以期望生成完全描绘真实图像的数据。这可以用于多种应用，例如图像去噪、绘画、结构化预测和探索强化学习。

这么做更深层次的愿景是，在构建生成模型的过程中，我们将使计算机理解这个世界以及它由哪些元素组成。

CHAPTER 7

第 7 章

深度信念网络

深度信念网络（Deep Belief Network, DBN）是一类由多层隐藏单元组成、层之间有连接的深度神经网络。DBN 的不同之处在于这些隐藏单元不与层内的其他单元交互。DBN 可以通过使用一组训练数据集来学习如何在无监督的情况下概率性地重建输入。它是大量随机变量相互作用下的联合（多变量）分布。DBN 是基于统计学与计算机科学的结合并依赖概率论、图算法、机器学习等概念的表达形式。

这些隐藏层可以充当特征检测器。经过训练后，DBN 可以通过监督训练来分类。

我们将在本章介绍以下内容：

- 理解深度信念网络。
- 模型训练。
- 标签预测。
- 调整模型的准确度。
- DBN 在 MNIST 数据集上的应用。
- DBN 中 RBM 层的神经元数量的影响。
- 具有两个 RBM 层的 DBN。
- 使用 DBN 对 NotMNIST 进行分类。

7.1 理解深度信念网络

DBN 可以看作由诸如**受限玻尔兹曼机（RBM）**或自编码器等简单的无监督网络构成。在这些网络中，每个子网络的隐藏层都服务于下一个可见层。RBM 是一个无向生成模型，具有（可见的）输入层和隐藏层，层之间有连接但连接不在层内。该拓扑结构会加速逐层无监督训练过程。对比散度应用于每个子网络，从最低层的数据集开始（最低可见层是训练数据集）。

DBN 采用一次训练一层的贪婪训练方式。这使得它成为第一个有效的深度学习算法之一。DBN 在实际场景中有着诸多的应用，我们来看一下如何使用 DBN 对 MNIST 和 NotMNIST 数据集进行分类。

DBN 实现

类实例化受限玻尔兹曼机层和损失函数。`DeepBeliefNetwork` 类自身就是 `Model` 的一个子类，见图 7-1。

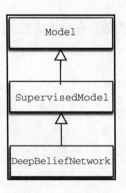

图 7-1 类的图示

1. 类初始化

在 DBN 初始化中，调用 `Model` 类的初始化方法 `_init_(self, name)`。`Model` 类属性信息如下：

- ❏ 输入数据：`self.input_data`。
- ❏ 数据标签：`self.input_labels`。
- ❏ 损失函数：`self.cost`。
- ❏ 最后一层节点数量：`self.layer_nodes`。
- ❏ TensorFlow 会话：`self.tf_session`。
- ❏ TensorFlow 图：`self.tf_graph= tf.graph`。

```
class Model(object):
    """Class representing an abstract Model."""

    def __init__(self, name):
        """Constructor.

        :param name: name of the model, used as filename.
            string, default 'dae'
        """
        self.name = name
        self.model_path = os.path.join(Config().models_dir, self.name)

        self.input_data = None
        self.input_labels = None
        self.keep_prob = None
        self.layer_nodes = []  # list of layers of the final network
        self.train_step = None
        self.cost = None

        # tensorflow objects
        self.tf_graph = tf.Graph()
        self.tf_session = None
        self.tf_saver = None
        self.tf_merged_summaries = None
        self.tf_summary_writer = None
```

定义的损失函数取值如下：

```
["cross_entropy", "softmax_cross_entropy", "mse"]
```

下面给出了 `DeepBeliefNetwork` 类的代码。如下所示，`_init_()` 函数声明了所有变量，例如指定每个 `RBM` 层的参数数组。还调用了 `SupervisedModel` 的 `_init_()` 函数，该函数是 `DeepBeliefNetwork` 类的超类。

两个重要参数的初始化方法如下：

- `self.rbms=[]`：RBM 类实例化数组。
- `self.rbm_graphs=[]`：每个 RBM 的 `tf.Graph` 数组。

```python
class DeepBeliefNetwork(SupervisedModel):
    """Implementation of Deep Belief Network for Supervised Learning.

    The interface of the class is sklearn-like.
    """

    def __init__(
        self, rbm_layers, name='dbn', do_pretrain=False,
        rbm_num_epochs=[10], rbm_gibbs_k=[1],
        rbm_gauss_visible=False, rbm_stddev=0.1, rbm_batch_size=[10],
        rbm_learning_rate=[0.01], finetune_dropout=1,
        finetune_loss_func='softmax_cross_entropy',
        finetune_act_func=tf.nn.sigmoid, finetune_opt='sgd',
        finetune_learning_rate=0.001, finetune_num_epochs=10,
            finetune_batch_size=20, momentum=0.5):
        SupervisedModel.__init__(self, name)

        self.loss_func = finetune_loss_func
        self.learning_rate = finetune_learning_rate
        self.opt = finetune_opt
        self.num_epochs = finetune_num_epochs
        self.batch_size = finetune_batch_size
        self.momentum = momentum
        self.dropout = finetune_dropout

        self.loss = Loss(self.loss_func)
        self.trainer = Trainer(
            finetune_opt, learning_rate=finetune_learning_rate,
            momentum=momentum)

        self.do_pretrain = do_pretrain
        self.layers = rbm_layers
        self.finetune_act_func = finetune_act_func

        # Model parameters
        self.encoding_w_ = []  # list of matrices of encoding weights per layer
        self.encoding_b_ = []  # list of arrays of encoding biases per layer

        self.softmax_W = None
        self.softmax_b = None

        rbm_params = {
            'num_epochs': rbm_num_epochs, 'gibbs_k': rbm_gibbs_k,
            'batch_size': rbm_batch_size, 'learning_rate':
        rbm_learning_rate}
```

```python
for p in rbm_params:
    if len(rbm_params[p]) != len(rbm_layers):
        # The current parameter is not specified by the user,
        # should default it for all the layers
        rbm_params[p] = [rbm_params[p][0] for _ in rbm_layers]

self.rbms = []
self.rbm_graphs = []

for l, layer in enumerate(rbm_layers):
    rbm_str = 'rbm-' + str(l+1)

    if l == 0 and rbm_gauss_visible:
        self.rbms.append(
            rbm.RBM(
                name=self.name + '-' + rbm_str,
                num_hidden=layer,
                learning_rate=rbm_params['learning_rate'][l],
                num_epochs=rbm_params['num_epochs'][l],
                batch_size=rbm_params['batch_size'][l],
                gibbs_sampling_steps=rbm_params['gibbs_k'][l],
                visible_unit_type='gauss', stddev=rbm_stddev))

    else:
        self.rbms.append(
            rbm.RBM(
                name=self.name + '-' + rbm_str,
                num_hidden=layer,
                learning_rate=rbm_params['learning_rate'][l],
                num_epochs=rbm_params['num_epochs'][l],
                batch_size=rbm_params['batch_size'][l],
                gibbs_sampling_steps=rbm_params['gibbs_k'][l]))

    self.rbm_graphs.append(tf.Graph())
```

注意一下 RBM 层是如何从 `rbm_layers` 数组构建的：

```python
for l, layer in enumerate(rbm_layers):
    rbm_str = 'rbm-' + str(l+1)

    if l == 0 and rbm_gauss_visible:
        self.rbms.append(
            rbm.RBM(
                name=self.name + '-' + rbm_str,
                num_hidden=layer,
                learning_rate=rbm_params['learning_rate'][l],
                num_epochs=rbm_params['num_epochs'][l],
                batch_size=rbm_params['batch_size'][l],
                gibbs_sampling_steps=rbm_params['gibbs_k'][l],
                visible_unit_type='gauss', stddev=rbm_stddev))
```

```
        else:
            self.rbms.append(
                rbm.RBM(
                    name=self.name + '-' + rbm_str,
                    num_hidden=layer,
                    learning_rate=rbm_params['learning_rate'][l],
                    num_epochs=rbm_params['num_epochs'][l],
                    batch_size=rbm_params['batch_size'][l],
                    gibbs_sampling_steps=rbm_params['gibbs_k'][l]))
```

2. RBM 类

对于每个 RBM 层，初始化一个 RBM 类。这个类扩展了 UnsupervisedModel 和 Model 类，见图 7-2。

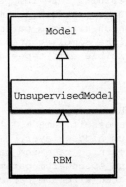

图 7-2 类的图示

以下代码给出了 RBM 类 _init_(..) 函数的详细信息：

```
class RBM(UnsupervisedModel):
    """Restricted Boltzmann Machine implementation using TensorFlow.

    The interface of the class is sklearn-like.
    """

    def __init__(
        self, num_hidden, visible_unit_type='bin',
        name='rbm', loss_func='mse', learning_rate=0.01,
        regcoef=5e-4, regtype='none', gibbs_sampling_steps=1,
            batch_size=10, num_epochs=10, stddev=0.1):
        """Constructor.

        :param num_hidden: number of hidden units
        :param loss_function: type of loss function
```

```
            :param visible_unit_type: type of the visible units (bin or gauss)
            :param gibbs_sampling_steps: optional, default 1
            :param stddev: default 0.1. Ignored if visible_unit_type is not
'gauss'
            """
            UnsupervisedModel.__init__(self, name)

            self.loss_func = loss_func
            self.learning_rate = learning_rate
            self.num_epochs = num_epochs
            self.batch_size = batch_size
            self.regtype = regtype
            self.regcoef = regcoef

            self.loss = Loss(self.loss_func)

            self.num_hidden = num_hidden
            self.visible_unit_type = visible_unit_type
            self.gibbs_sampling_steps = gibbs_sampling_steps
            self.stddev = stddev

            self.W = None
            self.bh_ = None
            self.bv_ = None

            self.w_upd8 = None
            self.bh_upd8 = None
            self.bv_upd8 = None

            self.cost = None

            self.input_data = None
            self.hrand = None
            self.vrand = None
```

一旦 RBM 图完成初始化,就会被追加到 TensorFlow 图中:

```
self.rbm_graphs.append(tf.Graph())
```

DBN 预训练

在本节中,我们看看如何进行 DBN 预训练:

```
    class RBM(UnsupervisedModel):
    ...
        def pretrain(self, train_set, validation_set=None):
            """Perform Unsupervised pretraining of the DBN."""
            self.do_pretrain = True
```

```python
    def set_params_func(rbmmachine, rbmgraph):
        params = rbmmachine.get_parameters(graph=rbmgraph)
        self.encoding_w_.append(params['W'])
        self.encoding_b_.append(params['bh_'])

    return SupervisedModel.pretrain_procedure(
        self, self.rbms, self.rbm_graphs, set_params_func=set_params_func,
        train_set=train_set, validation_set=validation_set)
```

回调 `SupervisedModel.pretrain_procedure(..)`，此类有以下参数：

- `layer_objs`：模型对象（自编码器或 RBM）列表。
- `layer_graphs`：`tf.Graph` 模型对象列表。
- `set_params_func`：预训练后用于设定参数的函数。
- `train_set`：训练数据集。
- `validation_set`：验证数据集。

该函数返回最后一层编码的数据：

```python
def pretrain_procedure(self, layer_objs, layer_graphs, set_params_func,
 train_set, validation_set=None):
    next_train = train_set
    next_valid = validation_set

    for l, layer_obj in enumerate(layer_objs):
        print('Training layer {}...'.format(l + 1))
        next_train, next_valid = self._pretrain_layer_and_gen_feed(
            layer_obj, set_params_func, next_train, next_valid,
            layer_graphs[l])

    return next_train, next_valid
```

回调 `self._pretrain_layer_and_gen_feed(...)`：

```python
def _pretrain_layer_and_gen_feed(self, layer_obj, set_params_func,
                                 train_set, validation_set, graph):
    layer_obj.fit(train_set, train_set,
                  validation_set, validation_set, graph=graph)

    with graph.as_default():
        set_params_func(layer_obj, graph)

        next_train = layer_obj.transform(train_set, graph=graph)
        if validation_set is not None:
            next_valid = layer_obj.transform(validation_set, graph=graph)
        else:
```

```
            next_valid = None
    return next_train, next_valid
```

在前面的函数中，`layer_obj` 被**迭代地**调用。

7.2 训练模型

使用 `fit()` 方法训练模型，它有以下参数：

- `train_X`: array_like, shape (n_samples, n_features)，训练集数据。
- `train_Y`: array_like, shape (n_samples, n_classes)，训练标签。
- `val_X`: array_like, shape (N, n_features) 可选, (default=None)，验证集数据。
- `val_Y`: array_like, shape (N, n_classes) 可选, (default=None)，验证集标签。
- `graph`: tf.Graph, optional (default=None)，TensorFlow 图对象。

接下来，我们看看如何用 `fit` 函数训练模型，并把模型保存在 `model_path` 指定的路径。

```
def fit(self, train_X, train_Y, val_X=None, val_Y=None, graph=None):

    if len(train_Y.shape) != 1:
        num_classes = train_Y.shape[1]
    else:
        raise Exception("Please convert the labels with one-hot encoding.")

    g = graph if graph is not None else self.tf_graph
    with g.as_default():
        # Build model
        self.build_model(train_X.shape[1], num_classes)
        with tf.Session() as self.tf_session:
            # Initialize tf stuff
            summary_objs = tf_utils.init_tf_ops(self.tf_session)
            self.tf_merged_summaries = summary_objs[0]
            self.tf_summary_writer = summary_objs[1]
            self.tf_saver = summary_objs[2]
            # Train model
            self._train_model(train_X, train_Y, val_X, val_Y)
```

```
            # Save model
            self.tf_saver.save(self.tf_session, self.model_path)
```

7.3 标签预测

可以通过调用以下方法来预测标签:

```
def predict(self, test_X):
    with self.tf_graph.as_default():
        with tf.Session() as self.tf_session:
            self.tf_saver.restore(self.tf_session, self.model_path)
            feed = {
                self.input_data: test_X,
                self.keep_prob: 1
            }
            return self.mod_y.eval(feed)
```

7.4 探索模型的准确度

模型的准确度可以通过计算测试集上的平均精确度来找到。实现方法如下:

```
def score(self, test_X, test_Y):
    ...
```

在这里,参数信息如下:

- `test_X: array_like, shape(n_samples, n_features)`,测试数据集。
- `test_Y: array_like, shape(n_samples, n_features)`,测试标签。
- `return float`:测试集上的平均准确度。

```
def score(self, test_X, test_Y):
    with self.tf_graph.as_default():
        with tf.Session() as self.tf_session:
            self.tf_saver.restore(self.tf_session, self.model_path)
            feed = {
                self.input_data: test_X,
                self.input_labels: test_Y,
                self.keep_prob: 1
            }
            return self.accuracy.eval(feed)
```

在下一节中,我们将看看如何把 DBN 实现应用在 MNIST 数据集上。

7.5 DBN 在 MNIST 数据集上的应用

让我们来看一下 DBN 类是如何最先在 MNIST 数据集上应用的。

7.5.1 加载数据集

首先,我们将 `idx3` 和 `idx1` 格式的数据集加载到测试集、训练集和验证集中。在这里,我们需要导入通用模块中定义的 TensorFlow 工具:

```
import tensorflow as tf
from common.models.boltzmann import dbn
from common.utils import datasets, utilities

trainX, trainY, validX, validY, testX, testY = 
    datasets.load_mnist_dataset(mode='supervised')
```

可以在以下代码中找到有关 `load_mnist_dataset()` 的详细信息。当设定 `mode='supervised'` 时,训练、测试和验证标签均会被返回:

```
def load_mnist_dataset(mode='supervised', one_hot=True):
    mnist = input_data.read_data_sets("MNIST_data/", one_hot=one_hot)
    # Training set
    trX = mnist.train.images
    trY = mnist.train.labels
    # Validation set
    vlX = mnist.validation.images
    vlY = mnist.validation.labels
    # Test set
    teX = mnist.test.images
 teY = mnist.test.labels
 if mode == 'supervised':
   return trX, trY, vlX, vlY, teX, teY
 elif mode == 'unsupervised':
   return trX, vlX, teX
```

7.5.2 具有 256 个神经元的 RBM 层的 DBN 的输入参数

我们将初始化前面定义 DBN 类所需的参数:

```
finetune_act_func = tf.nn.relu
rbm_layers = [256]
do_pretrain = True
name = 'dbn'
rbm_layers = [256]
finetune_act_func ='relu'
do_pretrain = True
rbm_learning_rate = [0.001]
rbm_num_epochs = [1]
rbm_gibbs_k= [1]
rbm_stddev= 0.1
rbm_gauss_visible= False
momentum= 0.5
rbm_batch_size= [32]
finetune_learning_rate = 0.01
finetune_num_epochs = 1
finetune_batch_size = 32
finetune_opt = 'momentum'
finetune_loss_func = 'softmax_cross_entropy'
finetune_dropout = 1
finetune_act_func = tf.nn.sigmoid
```

一旦完成参数定义，就可以在 MNIST 数据集上运行 DBN 网络：

```
srbm = dbn.DeepBeliefNetwork(
    name=name, do_pretrain=do_pretrain,
    rbm_layers=rbm_layers,
    finetune_act_func=finetune_act_func,
    rbm_learning_rate=rbm_learning_rate,
    rbm_num_epochs=rbm_num_epochs, rbm_gibbs_k = rbm_gibbs_k,
    rbm_gauss_visible=rbm_gauss_visible, rbm_stddev=rbm_stddev,
    momentum=momentum, rbm_batch_size=rbm_batch_size,
    finetune_learning_rate=finetune_learning_rate,
    finetune_num_epochs=finetune_num_epochs,
    finetune_batch_size=finetune_batch_size,
    finetune_opt=finetune_opt, finetune_loss_func=finetune_loss_func,
    finetune_dropout=finetune_dropout
    )

print(do_pretrain)
if do_pretrain:
    srbm.pretrain(trainX, validX)

# finetuning
print('Start deep belief net finetuning...')
srbm.fit(trainX, trainY, validX, validY)

# Test the model
print('Test set accuracy: {}'.format(srbm.score(testX, testY)))
```

7.5.3 具有 256 个神经元的 RBM 层的 DBN 的输出

上述代码的输出展示了测试数据集上的准确度：

```
Reconstruction loss: 0.156712: 100%|███████████| 5/5 [00:49<00:00, 9.99s/it]
Start deep belief net finetuning...
Tensorboard logs dir for this run is /home/ubuntu/.yadlt/logs/run53
Accuracy: 0.0868: 100%|███████████| 1/1 [00:04<00:00, 4.09s/it]
Test set accuracy: 0.0868000015616
```

整体的准确度和测试集的准确度非常低。随着迭代次数的增加，准确度有所改善。让我们用同样的样本运行 20 轮。

```
Reconstruction loss: 0.120337: 100%|███████████| 20/20 [03:07<00:00, 8.79s/it]
Start deep belief net finetuning...
Tensorboard logs dir for this run is /home/ubuntu/.yadlt/logs/run80
Accuracy: 0.105: 100%|███████████| 1/1 [00:04<00:00, 4.16s/it]
Test set accuracy: 0.10339999944
```

可以看出，损失有所下降，测试集准确度提高了 20%，达到 0.10339999944。

让我们把迭代周期（epoch）增加至 40，输出结果如下：

```
Reconstruction loss: 0.104798: 100%|███████████| 40/40 [06:20<00:00, 9.18s/it]
Start deep belief net finetuning...
Tensorboard logs dir for this run is /home/ubuntu/.yadlt/logs/run82
Accuracy: 0.075: 100%|███████████| 1/1 [00:04<00:00, 4.08s/it]
Test set accuracy: 0.0773999989033
As can be seen the accuracy again came down so the optimal number of
iterations peaks somewhere between 20 and 40
```

7.6 DBN 中 RBM 层的神经元数量的影响

我们来看看改变 RBM 层中的神经元数量将如何影响测试集的准确度。

7.6.1 具有 512 个神经元的 RBM 层

以下是在 RBM 层中具有 512 个神经元的 DBN 的输出。损失有所降低，测试集的准确度也降低了：

```
Reconstruction loss: 0.128517: 100%|███████████| 5/5 [01:32&lt;00:00,
19.25s/it]
Start deep belief net finetuning...
Tensorboard logs dir for this run is /home/ubuntu/.yadlt/logs/run55
Accuracy: 0.0758: 100%|███████████| 1/1 [00:06&lt;00:00, 6.40s/it]
Test set accuracy: 0.0689999982715
```

我们注意到损失和测试集准确度都下降了。这意味着增加神经元的数量并不一定会提高准确度。

7.6.2 具有 128 个神经元的 RBM 层

具有 128 个神经元的 RBM 层有更高的测试集准确度，但总体准确度较低：

```
Reconstruction loss: 0.180337: 100%|███████████| 5/5 [00:32&lt;00:00,
6.44s/it]
Start deep belief net finetuning...
Tensorboard logs dir for this run is /home/ubuntu/.yadlt/logs/run57
Accuracy: 0.0698: 100%|███████████| 1/1 [00:03&lt;00:00, 3.16s/it]
Test set accuracy: 0.0763999968767
```

7.6.3 准确度指标对比

由于我们已经在 RBM 层中训练了具有多个神经元数量的神经网络，让我们比较一下准确度。如图 7-3 所示，随着神经元数量的增加，重构损失有所下降。

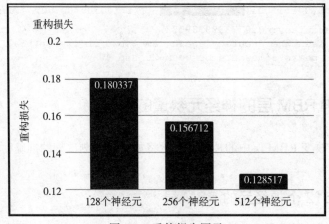

图 7-3 重构损失图示

具有 256 个神经元的测试集准确度达到峰值，然后下降，见图 7-4。

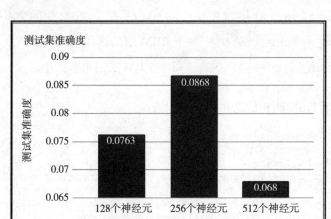

图 7-4　测试集准确度图示

7.7　具有两个 RBM 层的 DBN

在本节中，我们将创建一个具有两个 RBM 层的 DBN，并在 MNIST 数据集上实现。我们将修改 `DeepBeliefNetwork(..)` 类的输入参数：

```
name = 'dbn'
rbm_layers = [256, 256]
finetune_act_func ='relu'
do_pretrain = True
rbm_learning_rate = [0.001, 0.001]
rbm_num_epochs = [5, 5]
rbm_gibbs_k= [1, 1]
rbm_stddev= 0.1
rbm_gauss_visible= False
momentum= 0.5
rbm_batch_size= [32, 32]
finetune_learning_rate = 0.01
finetune_num_epochs = 1
finetune_batch_size = 32
finetune_opt = 'momentum'
finetune_loss_func = 'softmax_cross_entropy'
finetune_dropout = 1
finetune_act_func = tf.nn.sigmoid
```

请注意，一些参数有两个数组元素，所以我们需要为两个 RBM 层指定这些参数：

- rbm_layers=[256,256]：每个 RBM 层的神经元数量。
- rbm_learning_rate=[0.001,0001]：每个 RBM 层的学习率。
- rbm_num_epochs=[5,5]：每个 RBM 层的迭代周期数量。
- rbm_batch_size=[32,32]：每个 RBM 层的批量大小。

我们来看一下 DBN 的初始化和模型的训练：

```
srbm = dbn.DeepBeliefNetwork(
 name=name, do_pretrain=do_pretrain,
 rbm_layers=rbm_layers,
 finetune_act_func=finetune_act_func, rbm_learning_rate=rbm_learning_rate,
 rbm_num_epochs=rbm_num_epochs, rbm_gibbs_k = rbm_gibbs_k,
 rbm_gauss_visible=rbm_gauss_visible, rbm_stddev=rbm_stddev,
 momentum=momentum, rbm_batch_size=rbm_batch_size,
finetune_learning_rate=finetune_learning_rate,
 finetune_num_epochs=finetune_num_epochs,
finetune_batch_size=finetune_batch_size,
 finetune_opt=finetune_opt, finetune_loss_func=finetune_loss_func,
 finetune_dropout=finetune_dropout
 )

 if do_pretrain:
   srbm.pretrain(trainX, validX)

#
finetuning
 print('Start deep belief net finetuning...')
 srbm.fit(trainX, trainY, validX, validY)
```

测试模型信息：

```
print('Test set accuracy: {}'.format(srbm.score(testX, testY)))
```

完整的代码可在以下地址找到：https : //github.com/ml-resources/neuralnetwork-programming/blob/ed1/ch08/implementation/boltzmann/run_dbn_mnist_two_layers.py。

输出结果如下：

```
Reconstruction loss: 0.156286: 100%|████████| 5/5 [01:03&lt;00:00,
13.04s/it]
Training layer 2...
Tensorboard logs dir for this run is /home/ubuntu/.yadlt/logs/run73
Reconstruction loss: 0.127524: 100%|████████| 5/5 [00:23&lt;00:00,
```

```
4.87s/it]
Start deep belief net finetuning...
Tensorboard logs dir for this run is /home/ubuntu/.yadlt/logs/run74
Accuracy: 0.1496: 100%|███████████| 1/1 [00:05&lt;00:00, 5.53s/it]
Test set accuracy: 0.140300005674
```

从上面的输出信息可以看出，两个 RBM 层的测试集准确度比单个 RBM 层的 DBN 要好。

7.8 用 DBN 对 NotMNIST 数据集进行分类

让我们看一下 NotMNIST 数据集，该数据集在 2.5 节中有过探索，本节介绍 DBN 是如何在该数据集上工作的。

我们将利用与第 2 章中创建的相同的 pickle 文件（not MNIST.pickle）。初始化参数和导入方法如下：

```
import tensorflow as tf
import numpy as np
import cPickle as pickle

from common.models.boltzmann import dbn
from common.utils import datasets, utilities


flags = tf.app.flags
FLAGS = flags.FLAGS
pickle_file = '../notMNIST.pickle'

image_size = 28
num_of_labels = 10

RELU = 'RELU'
RELU6 = 'RELU6'
CRELU = 'CRELU'
SIGMOID = 'SIGMOID'
ELU = 'ELU'
SOFTPLUS = 'SOFTPLUS'
```

实现或多或少与 MNIST 数据集类似。主要实现代码如下：

```python
if __name__ == '__main__':
    utilities.random_seed_np_tf(-1)
    with open(pickle_file, 'rb') as f:
        save = pickle.load(f)
        training_dataset = save['train_dataset']
        training_labels = save['train_labels']
        validation_dataset = save['valid_dataset']
        validation_labels = save['valid_labels']
        test_dataset = save['test_dataset']
        test_labels = save['test_labels']
        del save  # hint to help gc free up memory
        print 'Training set', training_dataset.shape, training_labels.shape
        print 'Validation set', validation_dataset.shape, validation_labels.shape
        print 'Test set', test_dataset.shape, test_labels.shape

    train_dataset, train_labels = reformat(training_dataset, training_labels)
    valid_dataset, valid_labels = reformat(validation_dataset, validation_labels)
    test_dataset, test_labels = reformat(test_dataset, test_labels)

    #trainX, trainY, validX, validY, testX, testY = datasets.load_mnist_dataset(mode='supervised')
    trainX = train_dataset
    trainY = train_labels

    validX = valid_dataset
    validY = valid_labels
    testX = test_dataset
    testY = test_labels

    finetune_act_func = tf.nn.relu
    rbm_layers = [256]
    do_pretrain = True

    name = 'dbn'
    rbm_layers = [256]
    finetune_act_func ='relu'
    do_pretrain = True

    rbm_learning_rate = [0.001]

    rbm_num_epochs = [1]
    rbm_gibbs_k= [1]
    rbm_stddev= 0.1
    rbm_gauss_visible= False
    momentum= 0.5
    rbm_batch_size= [32]
    finetune_learning_rate = 0.01
    finetune_num_epochs = 1
    finetune_batch_size = 32
```

```
        finetune_opt = 'momentum'
        finetune_loss_func = 'softmax_cross_entropy'

        finetune_dropout = 1
        finetune_act_func = tf.nn.sigmoid
        srbm = dbn.DeepBeliefNetwork(
            name=name, do_pretrain=do_pretrain,
            rbm_layers=rbm_layers,
            finetune_act_func=finetune_act_func,
   rbm_learning_rate=rbm_learning_rate,
            rbm_num_epochs=rbm_num_epochs, rbm_gibbs_k = rbm_gibbs_k,
            rbm_gauss_visible=rbm_gauss_visible, rbm_stddev=rbm_stddev,
            momentum=momentum, rbm_batch_size=rbm_batch_size,
   finetune_learning_rate=finetune_learning_rate,
            finetune_num_epochs=finetune_num_epochs,
   finetune_batch_size=finetune_batch_size,
            finetune_opt=finetune_opt, finetune_loss_func=finetune_loss_fun
            finetune_dropout=finetune_dropout
        )

        if do_pretrain:
            srbm.pretrain(trainX, validX)

        # finetuning
        print('Start deep belief net finetuning...')
        srbm.fit(trainX, trainY, validX, validY)

        # Test the model
        print('Test set accuracy: {}'.format(srbm.score(testX, testY)))
```

> 完整代码可以从以下地址获取https://github.com/ml-resources/neuralnet work-programming/blob/ed1/ch08/implementation/boltzmann/run_dbn_nomnist.py。

上述代码的输出结果展示了我们的模型对于 NotMNIST 数据集的效果：

```
Reconstruction loss: 0.546223: 100%|          | 1/1 [00:00&lt;00:00, 5.51it/s]
Start deep belief net finetuning...
Tensorboard logs dir for this run is /home/ubuntu/.yadlt/logs/run76
Accuracy: 0.126: 100%|          | 1/1 [00:00&lt;00:00, 8.83it/s]
Test set accuracy: 0.180000007153
```

可以看出，这个网络比在 MNIST 数据集上表现得更好。

7.9 总结

在本章中,我们探索 DBN 并研究了如何使用一个或多个 RBM 层来构建分类管道。我们研究了 RBM 层内的各种参数及其对准确度、重构损失和测试集准确度的影响。我们还研究了使用一个或多个 RBM 的单层和多层 DBN。

在下一章,我们将研究生成模型以及它们和判别模型的区别。

CHAPTER 8

第 8 章

自 编 码 器

自编码器是一种神经网络,我们训练它使得它可以将其输入复制到其输出。它有一个隐藏层(我们称之为 h),隐藏层描述了表示输入的编码。网络可以看成由两部分组成:

- 编码函数:$h=f(x)$。
- 用于重建的解码函数:$r=g(h)$。

图 8-1 显示了一个带有 n 个输入和 m 个神经元的隐藏层的基本自编码器。

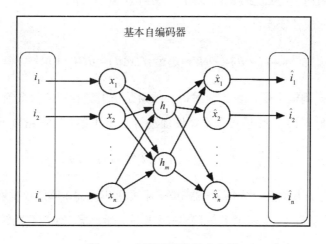

图 8-1 自编码器的基本表示

人们设计的自编码器无法学习对输入的完美复制。我们限制它们只能对输入进行近似的复制,并仅复制类似于训练数据的输入。由于模型被迫优先考虑输入的哪些部分应该被复制,它通常会学习数据的有用属性。

本章将涵盖以下内容:

- ❑ 自编码算法。
- ❑ 欠完备的自编码器。
- ❑ 基本自编码器。
- ❑ 加性高斯噪声自编码器。
- ❑ 稀疏自编码器。

8.1 自编码算法

在下面的表示中,x 是输入,y 是编码后的数据,z 是解码后的数据,σ 是非线性激活函数(通常为 sigmoid 或双曲正切),$f(x; \theta)$ 表示参数为 θ 的 x 的函数。

该模型可以以如下方式归纳。

将输入数据映射到隐藏层(进行编码)。映射通常是一个仿射(允许或保留平行关系)变换,然后进行非线性变换:

$$y=f(x; \theta)=\sigma(Wx+b) y=f(x; \theta)=\sigma(Wx+b)$$

将隐藏层映射到输出层,也称为**解码**。映射是仿射变换(仿射变换是保留点、直线和平面的线性映射方法),然后可以选择性地进行非线性变换。可由以下公式进行解释:

$$z=q(y; \theta')=g(f(x; \theta); \theta')=\sigma(W'y+b')$$

为了减小模型的大小,可以将编码器矩阵的权重和解码器矩阵的权重进行捆绑,这意味着我们会约束解码器权重矩阵,使得解码器权重矩阵是编码器权重矩阵的转置,即 $\theta'=\theta^{T}$。

隐藏层的维度可以比输入或输出层的维度高,也可以比它们低。

在隐藏层的维度比输入或输出层的维度低的情况下,解码器从较低维表示(也称为**欠**

完备表示）重建原始输入。为了整个算法正常工作，编码器应该学会提供那些能抓住数据本质（即分布变化的主要因素）的低维表示。因此，它必须找到一个表示数据的好方法。

 参考：http://blackecho.github.io/blog/machine-learning/2016/02/29/denoising-autoencoder-tensorflow.html。

8.2 欠完备自编码器

从自编码器获得有用特征的方法之一是通过将隐藏层 h 限制为比输入 x 更小的维度。编码层维度小于输入维度的自编码器称为欠完备自编码器。

学习不完备的表示会迫使自编码器捕捉训练数据的最显著特征。

学习过程为使损失函数 $L(x, g(f(x)))$ 最小化，其中 L 是损失函数，对于与 x 不相似的 $g(f(x))$，它会给出惩罚值，例如均方误差。

8.3 数据集

我们计划使用 `idx3` 格式的 MNIST 数据集作为输入来训练自编码器。我们使用前 100 张图像测试自编码器。先画出原始图像：

```
from tensorflow.examples.tutorials.mnist import input_data
import matplotlib.pyplot as plt


mnist = input_data.read_data_sets('MNIST_data', one_hot = True)

class OriginalImages:

    def __init__(self):
        pass

    def main(self):
        X_train, X_test = self.standard_scale(mnist.train.images,
mnist.test.images)

        original_imgs = X_test[:100]
        plt.figure(1, figsize=(10, 10))
```

```
        for i in range(0, 100):
            im = original_imgs[i].reshape((28, 28))
            ax = plt.subplot(10, 10, i + 1)
            for label in (ax.get_xticklabels() + ax.get_yticklabels()):
                label.set_fontsize(8)

            plt.imshow(im, cmap="gray", clim=(0.0, 1.0))
        plt.suptitle(' Original Images', fontsize=15, y=0.95)
        plt.savefig('figures/original_images.png')
        plt.show()


def main():
    auto = OriginalImages()
    auto.main()

if __name__ == '__main__':
    main()
```

输出如图 8-2 所示。

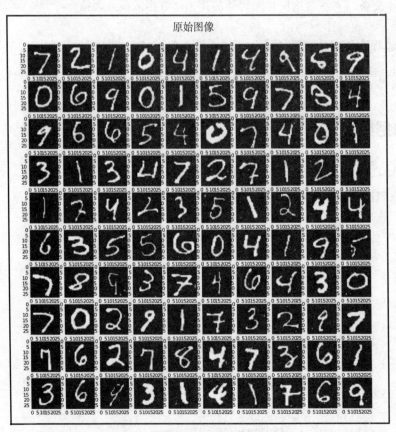

图 8-2　原始 MNIST 图像的绘制

8.4 基本自编码器

我们看一个自编码器的基本例子，它同时也是基本的自编码器。

首先，创建一个 `AutoEncoder` 类，将以下参数传给 `__init__()` 函数进行初始化：

- `num_input`：输入样本数量。
- `num_hidden`：隐藏层神经元数量。
- `transfer_function = tf.nn.softplus`：变换函数。
- `optimizer = tf.train.AdamOptimizer()`：优化方法。

你可以传入自定义的 `transfer_function` 和 `optimizer`，或者使用指定的函数。在本例中，我们使用 `softplus` 作为默认的 `transfer_function`（也称为激活函数）：$f(x)=\ln(1+e^x)$。

8.4.1 自编码器的初始化

首先，我们初始化类变量和权重：

```
self.num_input = num_input
self.num_hidden = num_hidden
self.transfer = transfer_function
network_weights = self._initialize_weights()
self.weights = network_weights
```

在这里，`_initialize_weigths()` 函数是一个用于初始化权重字典的值的局部函数：

- `w1` 是大小为 `num_input`×`num_hidden` 的二维张量。
- `b1` 是大小为 `num_hidden` 的一维张量。
- `w2` 是大小为 `num_hidden`×`num_input` 的二维张量。
- `b2` 是大小为 `num_input` 的一维张量。

以下代码显示了如何以 TensorFlow 变量字典的形式初始化两个隐藏层的权重：

```
def _initialize_weights(self):
    weights = dict()
    weights['w1'] = tf.get_variable("w1", shape=[self.num_input, self.num_hidden],
                    initializer=tf.contrib.layers.xavier_initializer())
    weights['b1'] = tf.Variable(tf.zeros([self.num_hidden], dtype=tf.float32))
    weights['w2'] = tf.Variable(tf.zeros([self.num_hidden, self.num_input],
        dtype=tf.float32))
    weights['b2'] = tf.Variable(tf.zeros([self.num_input], dtype=tf.float32))
    return weights
```

接下来我们定义 `x_var`、`hidden_layer` 和 `reconstruction` 层：

```
self.x_var = tf.placeholder(tf.float32, [None, self.num_input])
self.hidden_layer = self.transfer(tf.add(tf.matmul(self.x_var,
    self.weights['w1']), self.weights['b1']))
self.reconstruction = tf.add(tf.matmul(self.hidden_layer,
    self.weights['w2']), self.weights['b2'])
```

```
This is followed by the cost function and the Optimizer
# cost function
self.cost = 0.5 * tf.reduce_sum(
    tf.pow(tf.subtract(self.reconstruction, self.x_var), 2.0))
self.optimizer = optimizer.minimize(self.cost)
```

$$\text{cost} = \frac{1}{2}\sqrt{\sum_{0}^{m-1}\sum_{0}^{n-1}(\hat{x}-x)^2}$$

实例化全局变量初始化函数，将其传入 TensorFlow 会话。

```
initializer = tf.global_variables_initializer()
self.session = tf.Session()
self.session.run(initializer)
```

8.4.2　AutoEncoder 类

以下代码显示了 `AutoEncoder` 类。我们将在下面几节实例化该类，以构建自编码器：

```
import tensorflow as tf

class AutoEncoder:
```

```python
    def __init__(self, num_input, num_hidden,
        transfer_function=tf.nn.softplus,
   optimizer = tf.train.AdamOptimizer()):
      self.num_input = num_input
      self.num_hidden = num_hidden
      self.transfer = transfer_function

      network_weights = self._initialize_weights()
      self.weights = network_weights

      # model for reconstruction of the image
      self.x_var = tf.placeholder(tf.float32, [None, self.num_input])
      self.hidden_layer = self.transfer(tf.add(tf.matmul(self.x_var,
          self.weights['w1']), self.weights['b1']))
      self.reconstruction = tf.add(tf.matmul(self.hidden_layer,
          self.weights['w2']), self.weights['b2'])

      # cost function
      self.cost =
          0.5 * tf.reduce_sum(tf.pow(tf.subtract(self.reconstruction,
          self.x_var), 2.0))
      self.optimizer = optimizer.minimize(self.cost)

      initializer = tf.global_variables_initializer()
      self.session = tf.Session()
      self.session.run(initializer)

   def _initialize_weights(self):
      weights = dict()
      weights['w1'] = tf.get_variable("w1",
                        shape=[self.num_input,
                        self.num_hidden],
                        initializer=
                            tf.contrib.layers.xavier_initializer())
      weights['b1'] = tf.Variable(tf.zeros([self.num_hidden],
                            dtype=tf.float32))
      weights['w2'] = tf.Variable(
         tf.zeros([self.num_hidden, self.num_input],
         dtype=tf.float32))
      weights['b2'] = tf.Variable(tf.zeros(
                        [self.num_input], dtype=tf.float32))
      return weights

   def partial_fit(self, X):
      cost, opt = self.session.run((self.cost, self.optimizer),
         feed_dict={self.x_var: X})
      return cost

   def calculate_total_cost(self, X):
      return self.session.run(self.cost, feed_dict = {self.x_var: X})
```

```python
    def transform(self, X):
        return self.session.run(self.hidden_layer,
            feed_dict={self.x_var: X})

    def generate(self, hidden = None):
        if hidden is None:
            hidden = self.session.run(
                tf.random_normal([1, self.num_hidden]))
        return self.session.run(self.reconstruction,
                feed_dict={self.hidden_layer: hidden})

    def reconstruct(self, X):
        return self.session.run(self.reconstruction,
                        feed_dict={self.x_var: X})

    def get_weights(self):
        return self.session.run(self.weights['w1'])

    def get_biases(self):
        return self.session.run(self.weights['b1'])
```

8.4.3 应用于 MNIST 数据集的基本自编码器

将基本自编码器应用于 MNIST 数据集：

`input_data.read_data_sets('MNIST_data', one_hot = True)`。

使用 Scikit Learn 的 `sklearn.preprocessing` 模块中的 `StandardScalar` 提取 `testmnist.test.images` 和训练图片 `mnist.train.images`：

```
X_train, X_test = self.standard_scale(mnist.train.images,
mnist.test.images).
```

> 预处理模块提供了一个实用的类 `StandardScaler`，它提供了数据转换的接口。它计算了训练集的平均值和标准差然后对训练集做了变换。它对测试集做了相同的变换。默认情况下，这种变换会对数据进行中心化（均值为 0）并使方差为 1。
> 可以通过将 `with_mean = False` 或 `with_std = False` 传递给 `StandardScaler` 的构造函数来禁用去均值或对数据进行缩放。

接下来，定义一个之前提到的 `AutoEncoder` 类的实例：

```
n_samples = int(mnist.train.num_examples)
training_epochs = 5
batch_size = 128
display_step = 1

autoencoder = AutoEncoder(n_input = 784,
  n_hidden = 200,
  transfer_function = tf.nn.softplus,
  optimizer = tf.train.AdamOptimizer(learning_rate = 0.001))
```

注意，该自编码器包含如下元素：

- 输入神经元数量 784。
- 隐藏层神经元数量 200。
- 激活函数是 `tf.nn.softplus`。
- 优化方法是 `tf.train.AdamOptimizer`。

接下来，我们迭代训练数据并展示损失函数：

```
for epoch in range(training_epochs):
    avg_cost = 0.
    total_batch = int(n_samples / batch_size)
    # Loop over all batches
    for i in range(total_batch):
        batch_xs = self.get_random_block_from_data(X_train, batch_size)

        # Fit training using batch data
        cost = autoencoder.partial_fit(batch_xs)
        # Compute average loss
        avg_cost += cost / n_samples * batch_size

    # Display logs per epoch step
    if epoch % display_step == 0:
        print("Epoch:", '%04d' % (epoch + 1), "cost=",
"{:.9f}".format(avg_cost))
```

打印总的损失：

```
print("Total cost: " + str(autoencoder.calc_total_cost(X_test)))
```

每个迭代周期的输出如下；和期望的一样，随着迭代次数的增加，损失会收敛。

```
('Epoch:', '0001', 'cost=', '20432.278386364')
('Epoch:', '0002', 'cost=', '13542.435997727')
('Epoch:', '0003', 'cost=', '10630.662196023')
('Epoch:', '0004', 'cost=', '10717.897946591')
('Epoch:', '0005', 'cost=', '9354.191921023')
Total cost: 824850.0
```

1. 绘制基本自编码器的权重

训练完毕后,使用 Matplotlib 库绘制权重(见图 8-3),代码如下所示:

```
print("Total cost: " + str(autoencoder.calc_total_cost(X_test)))
wts = autoencoder.getWeights()
dim = math.ceil(math.sqrt(autoencoder.n_hidden))
plt.figure(1, figsize=(dim, dim))

for i in range(0, autoencoder.n_hidden):
    im = wts.flatten()[i::autoencoder.n_hidden].reshape((28, 28))
    ax = plt.subplot(dim, dim, i + 1)
    for label in (ax.get_xticklabels() + ax.get_yticklabels()):
        label.set_fontname('Arial')
        label.set_fontsize(8)
    # plt.title('Feature Weights ' + str(i))
    plt.imshow(im, cmap="gray", clim=(-1.0, 1.0))
plt.suptitle('Basic AutoEncoder Weights', fontsize=15, y=0.95)
#plt.title("Test Title", y=1.05)
plt.savefig('figures/basic_autoencoder_weights.png')
plt.show()
```

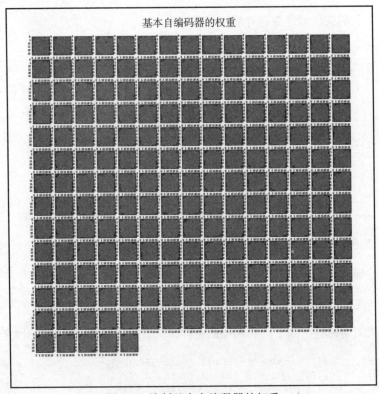

图 8-3 绘制基本自编码器的权重

在下一节中,我们将看到如何使用图 8-3 中显示的权重来重建图像。

2. 绘制基本自编码器重建的图像

重建图像后,我们将它们展示出来,了解它们的样子。首先,我们将使用之前创建的 autoencoder 实例重建图像:

```
predicted_imgs = autoencoder.reconstruct(X_test[:100])

plt.figure(1, figsize=(10, 10))
for i in range(0, 100):
    im = predicted_imgs[i].reshape((28, 28))
    ax = plt.subplot(10, 10, i + 1)
    for label in (ax.get_xticklabels() + ax.get_yticklabels()):
            label.set_fontname('Arial')
            label.set_fontsize(8)

    plt.imshow(im, cmap="gray", clim=(0.0, 1.0))
plt.suptitle('Basic AutoEncoder Images', fontsize=15, y=0.95)
plt.savefig('figures/basic_autoencoder_images.png')
plt.show()
```

神经网络创建的图像如图 8-4 所示。

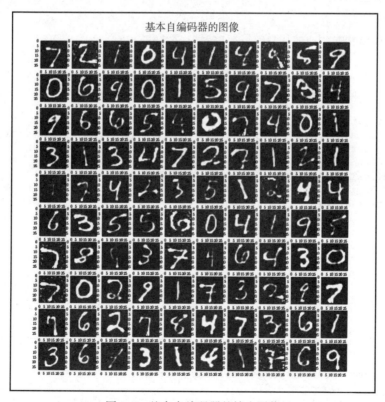

图 8-4 基本自编码器的输出图像

8.4.4 基本自编码器的完整代码

这里有完整代码，或者也可以从下面的 GitHub 地址下载：https://github.com/rajdeepd/neuralnetwork-programming/blob/ed1/ch07/basic_autoencoder_example.py。

```
import numpy as np

import sklearn.preprocessing as prep
import tensorflow as tf
from tensorflow.examples.tutorials.mnist import input_data
from autencoder_models.auto_encoder import AutoEncoder
import math
import matplotlib.pyplot as plt


mnist = input_data.read_data_sets('MNIST_data', one_hot = True)

class BasicAutoEncoder:

    def __init__(self):
        pass

    def standard_scale(self,X_train, X_test):
        preprocessor = prep.StandardScaler().fit(X_train)
        X_train = preprocessor.transform(X_train)
        X_test = preprocessor.transform(X_test)
        return X_train, X_test

    def get_random_block_from_data(self,data, batch_size):
        start_index = np.random.randint(0, len(data) - batch_size)
        return data[start_index:(start_index + batch_size)]

    def main(self):
        X_train, X_test = self.standard_scale(mnist.train.images,
mnist.test.images)

        n_samples = int(mnist.train.num_examples)
        training_epochs = 5
        batch_size = 128
        display_step = 1

        autoencoder = AutoEncoder(n_input = 784,
                                  n_hidden = 200,
                                  transfer_function = tf.nn.softplus,
                                  optimizer = tf.train.AdamOptimizer(
                                      learning_rate = 0.001))

        for epoch in range(training_epochs):
            avg_cost = 0.
            total_batch = int(n_samples / batch_size)
```

```python
            # Loop over all batches
            for i in range(total_batch):
                batch_xs = self.get_random_block_from_data(X_train,
batch_size)

                # Fit training using batch data
                cost = autoencoder.partial_fit(batch_xs)
                # Compute average loss
                avg_cost += cost / n_samples * batch_size

            # Display logs per epoch step
            if epoch % display_step == 0:
                print("Epoch:", '%04d' % (epoch + 1), "cost=",
"{:.9f}".format(avg_cost))

        print("Total cost: " + str(autoencoder.calc_total_cost(X_test)))

        wts = autoencoder.getWeights()
        dim = math.ceil(math.sqrt(autoencoder.n_hidden))
        plt.figure(1, figsize=(dim, dim))

        for i in range(0, autoencoder.n_hidden):
            im = wts.flatten()[i::autoencoder.n_hidden].reshape((28, 28))
            ax = plt.subplot(dim, dim, i + 1)
            for label in (ax.get_xticklabels() + ax.get_yticklabels()):
                label.set_fontname('Arial')
                label.set_fontsize(8)
            # plt.title('Feature Weights ' + str(i))
            plt.imshow(im, cmap="gray", clim=(-1.0, 1.0))
        plt.suptitle('Basic AutoEncoder Weights', fontsize=15, y=0.95)
        #plt.title("Test Title", y=1.05)
        plt.savefig('figures/basic_autoencoder_weights.png')
        plt.show()

        predicted_imgs = autoencoder.reconstruct(X_test[:100])

        plt.figure(1, figsize=(10, 10))

        for i in range(0, 100):
            im = predicted_imgs[i].reshape((28, 28))
            ax = plt.subplot(10, 10, i + 1)
            for label in (ax.get_xticklabels() + ax.get_yticklabels()):
                label.set_fontname('Arial')
                label.set_fontsize(8)
            plt.imshow(im, cmap="gray", clim=(0.0, 1.0))
        plt.suptitle('Basic AutoEncoder Images', fontsize=15, y=0.95)
        plt.savefig('figures/basic_autoencoder_images.png')
        plt.show()

def main():
    auto = BasicAutoEncoder()
```

```
    auto.main()

if __name__ == '__main__':
    main()
```

8.4.5 基本自编码器小结

自编码器使用 200 个神经元的隐藏层构建了 MNSIT 图像的基本近似。图 8-5 显示了 9 个图像以及如何使用基本自编码器将它们转换为近似值:

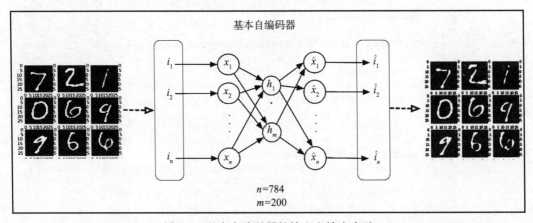

图 8-5　基本自编码器的输入和输出表示

在下一节中,我们会看一个更高级的自编码器——**加性高斯噪声自编码器**。

8.5　加性高斯噪声自编码器

什么是降噪自编码器?它们与我们在前面小节中看到的基本模型非常相似,区别在于,输入在传递到网络之前已被破坏。在训练时将原始输入(不是已损坏的输入)与重建的输出进行比较,该自编码器会从损坏的图像中重建原始输入图像。从损坏的图像重建原始图像的能力使得自编码器非常智能。

加性噪声自编码器使用以下等式来对输入数据进行破坏:

$$x_{corr} = x + scale \times random\_normal(n)$$

以下是关于上述等式的详细描述：

- x 是原始图像。
- scale 是从 n 中生成的随机正态数的乘子。
- n 是训练样本数。
- x_{corr} 是被破坏之后的输出。

8.5.1 自编码器类

我们通过传递以下参数初始化 `AdditiveGaussianNoiseAutoEncoder` 类中定义的自编码器：

- `num_input`：输入样本数量。
- `num_hidden`：隐藏层神经元数量。
- `transfer_function = tf.nn.sigmoid`：变换函数。
- `optimizer = tf.train.AdamOptimizer()`：优化方法。
- `scale = 0.1`：图像破坏的程度。

```
def __init__(self, num_input, num_hidden,
             transfer_function=tf.nn.sigmoid,
             optimizer=tf.train.AdamOptimizer(),
             scale=0.1):
```

将传递的参数赋给实例变量：

```
self.num_input = num_input
self.num_hidden = num_hidden
self.transfer = transfer_function
self.scale = tf.placeholder(tf.float32)
self.training_scale = scale
n_weights = self._initialize_weights()
self.weights = n_weights
```

初始化隐藏层 `hidden_layer` 和重建层 `reconstruction`：

```
self.x = tf.placeholder(tf.float32, [None, self.num_input])
self.hidden_layer = self.transfer(
        tf.add(tf.matmul(
                self.x + scale * tf.random_normal((n_input,)),
                self.weights['w1']),
```

```
                    self.weights['b1']))
self.reconstruction = tf.add(
                    tf.matmul(self.hidden_layer, self.weights['w2']),
                    self.weights['b2'])
```

定义损失函数和优化方法：

```
self.cost = 0.5 * tf.reduce_sum(tf.pow(tf.subtract(
                    self.reconstruction, self.x), 2.0))
self.optimizer = optimizer.minimize(self.cost)
```

加性高斯自编码器的损失函数和基本自编码器一样：

$$\text{cost} = \frac{1}{2}\sqrt{\sum_{0}^{m-1}\sum_{0}^{n-1}(\hat{x}-x)^2}$$

最后，初始化全局变量，创建 TensorFlow 会话，执行 init：

```
init = tf.global_variables_initializer()
self.session = tf.Session()
self.session.run(init)
```

在下一节，会介绍自编码器如何对 MNIST 数据集进行编码。

8.5.2 应用于 MNIST 数据集的加性高斯自编码器

首先，读取训练 X_train 和测试数据 X_test：

```
mnist = input_data.read_data_sets('MNIST_data', one_hot=True)

def get_random_block_from_data(data, batch_size):
    start_index = np.random.randint(0, len(data) - batch_size)
    return data[start_index:(start_index + batch_size)]

X_train = mnist.train.images
X_test = mnist.test.images
```

定义样本数量的变量 n_samples、training_epoch，对每次迭代和 display_step 定义 batch_size：

```
n_samples = int(mnist.train.num_examples)
training_epochs = 2
batch_size = 128
display_step = 1
```

实例化自编码器和优化方法。该自编码器的隐藏层有200个神经元，使用sigmoid作为`transfer_function`：

```
autoencoder = AdditiveGaussianNoiseAutoEncoder(n_input=784,
                                               n_hidden=200,
transfer_function=tf.nn.sigmoid,
optimizer=tf.train.AdamOptimizer(learning_rate=0.001),
                                               scale=0.01)
```

1. 训练模型

一旦定义了神经网络层，对每个`batch`的数据，我们通过调用方法`autoencoder.partial_fit(batch_xs)`训练模型：

```
for epoch in range(training_epochs):
    avg_cost = 0.
    total_batch = int(n_samples / batch_size)
    # Loop over all batches
    for i in range(total_batch):
        batch_xs = get_random_block_from_data(X_train, batch_size)

        # Fit training using batch data
        cost = autoencoder.partial_fit(batch_xs)
        # Compute average loss
        avg_cost += cost / n_samples * batch_size

    # Display logs per epoch step
    if epoch % display_step == 0:
        print("Epoch:", '%04d' % (epoch + 1), "cost=", avg_cost)

print("Total cost: " + str(autoencoder.calc_total_cost(X_test)))
```

每个迭代周期的损失如下：

```
('Epoch:', '0001', 'cost=', 1759.873304261363)
('Epoch:', '0002', 'cost=', 686.85984829545475)
('Epoch:', '0003', 'cost=', 460.52834446022746)
('Epoch:', '0004', 'cost=', 355.10590241477308)
('Epoch:', '0005', 'cost=', 297.99104825994351)
```

训练集总的损失如下：

```
Total cost: 21755.4
```

2. 绘制权重图

用Matplotlib直观地绘制权重图（见图8-6）：

```python
wts = autoencoder.get_weights()
dim = math.ceil(math.sqrt(autoencoder.num_hidden))
plt.figure(1, figsize=(dim, dim))
for i in range(0, autoencoder.num_hidden):
    im = wts.flatten()[i::autoencoder.num_hidden].reshape((28, 28))
    ax = plt.subplot(dim, dim, i + 1)
    for label in (ax.get_xticklabels() + ax.get_yticklabels()):
        label.set_fontsize(8)
    #plt.title('Feature Weights ' + str(i))

    plt.imshow(im, cmap="gray", clim=(-1.0, 1.0))
plt.suptitle('Additive Gaussian Noise AutoEncoder Weights', fontsize=15,
y=0.95)
plt.savefig('figures/additive_gaussian_weights.png')
plt.show()
```

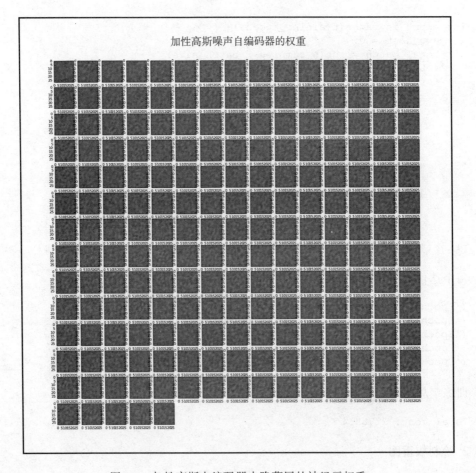

图 8-6　加性高斯自编码器中隐藏层的神经元权重

8.5.3 绘制重建的图像

最后一步是绘制重建后的图像,以便让我们看到编码器如何根据权重重建图像(见图 8-7):

```
predicted_imgs = autoencoder.reconstruct(X_test[:100])

# plot the reconstructed images
plt.figure(1, figsize=(10, 10))
plt.title('Autoencoded Images')
for i in range(0, 100):
    im = predicted_imgs[i].reshape((28, 28))
    ax = plt.subplot(10, 10, i + 1)
    for label in (ax.get_xticklabels() + ax.get_yticklabels()):
        label.set_fontname('Arial')
        label.set_fontsize(8)

    plt.imshow(im, cmap="gray", clim=(0.0, 1.0))
plt.suptitle('Additive Gaussian Noise AutoEncoder Images', fontsize=15,
y=0.95)
plt.savefig('figures/additive_gaussian_images.png')
plt.show()
```

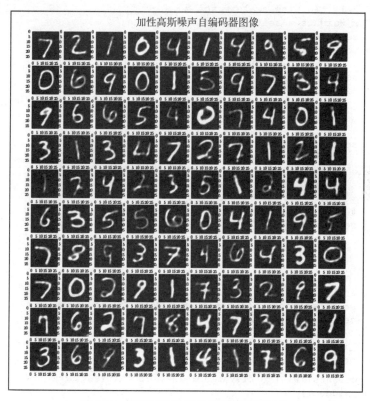

图 8-7 使用加性高斯自编码器重建图像

8.5.4　加性高斯自编码器的完整代码

以下是加性高斯自编码器的完整代码：

```
import numpy as np
import tensorflow as tf
def xavier_init(fan_in, fan_out, constant = 1):
    low = -constant * np.sqrt(6.0 / (fan_in + fan_out))
    high = constant * np.sqrt(6.0 / (fan_in + fan_out))
    return tf.random_uniform((fan_in, fan_out), minval = low, maxval = high, dtype = tf.float32)

class AdditiveGaussianNoiseAutoEncoder(object):
    def __init__(self, num_input, num_hidden,
                 transfer_function=tf.nn.sigmoid,
                 optimizer=tf.train.AdamOptimizer(),
                 scale=0.1):
        self.num_input = num_input
        self.num_hidden = num_hidden
        self.transfer = transfer_function
        self.scale = tf.placeholder(tf.float32)
        self.training_scale = scale
        n_weights = self._initialize_weights()
        self.weights = n_weights

        # model
        self.x = tf.placeholder(tf.float32, [None, self.num_input])
        self.hidden_layer = self.transfer(
            tf.add(tf.matmul(
                    self.x + scale * tf.random_normal((n_input,)),
                    self.weights['w1']),
                    self.weights['b1']))
        self.reconstruction = tf.add(
                        tf.matmul(self.hidden_layer, self.weights['w2']),
                                self.weights['b2'])

        # cost
        self.cost = 0.5 * tf.reduce_sum(tf.pow(tf.subtract(
                        self.reconstruction, self.x), 2.0))

        self.optimizer = optimizer.minimize(self.cost)

        init = tf.global_variables_initializer()
        self.session = tf.Session()
        self.session.run(init)

    def _initialize_weights(self):
        weights = dict()
        weights['w1'] = tf.Variable(xavier_init(self.num_input, self.num_hidden))
        weights['b1'] = tf.Variable(tf.zeros([self.num_hidden],
```

```
            dtype=tf.float32))
            weights['w2'] = tf.Variable(tf.zeros([self.num_hidden,
self.num_input],
            dtype=tf.float32))
            weights['b2'] = tf.Variable(tf.zeros([self.num_input],
dtype=tf.float32))
        return weights

    def partial_fit(self, X):
        cost, opt = self.session.run((self.cost, self.optimizer),
            feed_dict={self.x: X,self.scale: self.training_scale})
        return cost

    def kl_divergence(self, p, p_hat):
        return tf.reduce_mean(
          p * tf.log(p) - p * tf.log(p_hat) + (1 - p) * tf.log(1 - p) - (1 - p) *      tf.log(1 - p_hat))

    def calculate_total_cost(self, X):
        return self.session.run(self.cost, feed_dict={self.x: X,
                                                      self.scale:
self.training_scale
                                                       })

    def transform(self, X):
        return self.session.run(
          self.hidden_layer,
          feed_dict = {self.x: X, self.scale: self.training_scale})

    def generate_value(self, _hidden=None):
        if _hidden is None:
            _hidden = np.random.normal(size=self.weights["b1"])
        return self.session.run(self.reconstruction,
            feed_dict={self.hidden_layer: _hidden})

    def reconstruct(self, X):
        return self.session.run(self.reconstruction,
          feed_dict={self.x: X,self.scale: self.training_scale })

    def get_weights(self):
        return self.session.run(self.weights['w1'])

    def get_biases(self):
        return self.session.run(self.weights['b1'])
```

8.5.5 比较基本自编码器和加性高斯噪声自编码器

图 8-8 展示了每个迭代周期两个算法的损失值。可以看到，基本自编码器的损失比加性高斯自编码器的损失高很多。

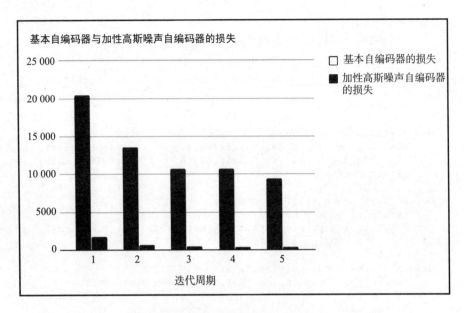

图 8-8　基本自编码器与加性高斯噪声自编码器的损失比较

8.5.6　加性高斯噪声自编码器小结

你知道了如何构建具有高斯噪声的自编码器,与基本自编码器相比,它可以大幅提高模型的准确性。

8.6　稀疏自编码器

在本节中,我们将看看如何将稀疏性加到损失函数中,这有助于降低训练的损失(值)。大部分代码保持不变,主要变化为计算损失函数的方式。

8.6.1　KL 散度

我们首先理解 KL 散度,它用于在损失函数中引入稀疏度元素。

如果神经元的输出值接近于 1,我们可以将神经元视为激活(或触发),如果其输出值接近于零,则可以将其视为非激活。我们希望将神经元大部分时间内都限制在非激活状态。以上论述假定了激活函数为 sigmoid 函数。

$a_j^{(2)}$ 表示隐藏层第 j 个神经元的激活状态。这一表达式并没有明确说明是哪个输入 x 导致了神经元的激活状态。我们用 $a_j^{(2)}(x)$ 表示对于一个特定的输入 x，隐藏层神经元 j 的激活状态。

进一步，我们用 $\hat{\rho}_j = \dfrac{1}{m} \sum_{i=1}^{m} \left[a_j^{(2)}(x^{(i)}) \right]$ 表示隐藏层神经元 j 的平均激活状态（在训练集上取平均）。我们想要有个大概的约束 $\hat{\rho}_j = \rho$，其中 ρ 是稀疏参数，通常是一个接近于零的值（比如，0.05）。我们的目标是使每个隐藏的神经元 j 的平均激活状态接近 0.05。为了满足前面的约束，隐藏层神经元的激活状态必须大部分接近于零。

为了达到这个目的，需要在优化目标中加入额外的惩罚项，它对明显偏离 ρ 的神经元进行惩罚：

$$\sum_{j=1}^{s_2} \rho \log \frac{\rho}{\hat{\rho}_j} + (1-\rho)\log \frac{1-\rho}{1-\hat{\rho}_j}$$

图 8-9 展示了 KL 散度如何随平均激活状态的变化而变化。

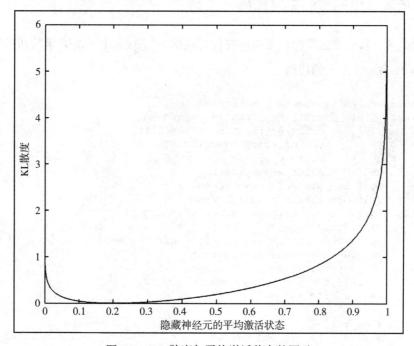

图 8-9　KL 散度与平均激活状态的图示

TensorFlow 中的 KL 散度

在我们实现的稀疏编码器中，我们在 SparseEncoder 类的 kl_divergence 函数中定义了 KL 散度，这只是上述公式的实现：

```
def kl_divergence(self, p, p_hat):
    return tf.reduce_mean(
        p*(tf.log(p)/tf.log(p_hat)) +
        (1-p)*(tf.log(1-p)/tf.log(1-p_hat)))
```

基于 KL 散度的稀疏自编码器的损失

与本章之前讨论的编码器相比，使用两个新参数 sparse_reg 和 kl_divergence 重新定义了损失函数：

```
self.cost = 0.5 * tf.reduce_sum(
  tf.pow(tf.subtract(self.reconstruction, self.x), 2.0)) +
    self.sparse_reg * self.kl_divergence(self.sparsity_level,
self.hidden_layer)
```

8.6.2 稀疏自编码器的完整代码

作为参考，我们在这里给出了带有先前讨论的 kl_divergence 函数和损失函数的 SparseAutoEncoder 的代码：

```
class SparseAutoencoder(object):
    def __init__(self, num_input, num_hidden,
                transfer_function=tf.nn.softplus,
                optimizer=tf.train.AdamOptimizer(),
                scale=0.1):
        self.num_input = num_input
        self.num_hidden = num_hidden
        self.transfer = transfer_function
        self.scale = tf.placeholder(tf.float32)
        self.training_scale = scale
        network_weights = self._initialize_weights()
        self.weights = network_weights
        self.sparsity_level = np.repeat([0.05],
            self.num_hidden).astype(np.float32)
        self.sparse_reg = 0.0

        # model
        self.x = tf.placeholder(tf.float32, [None, self.num_input])
        self.hidden_layer = self.transfer(tf.add(tf.matmul(
```

```python
            self.x + scale * tf.random_normal((num_input,)),
                                    self.weights['w1']),
                                    self.weights['b1']))
        self.reconstruction = tf.add(tf.matmul(self.hidden_layer,
            self.weights['w2']), self.weights['b2'])

        # cost
        self.cost = 0.5 * tf.reduce_sum(
            tf.pow(tf.subtract(self.reconstruction, self.x), 2.0)) +
            self.sparse_reg * self.kl_divergence(
                self.sparsity_level, self.hidden_layer)
        self.optimizer = optimizer.minimize(self.cost)

        init = tf.global_variables_initializer()
        self.session = tf.Session()
        self.session.run(init)

    def _initialize_weights(self):
        all_weights = dict()
        all_weights['w1'] = tf.Variable(xavier_init(self.num_input,
            self.num_hidden))
        all_weights['b1'] = tf.Variable(tf.zeros([self.num_hidden],
            dtype = tf.float32))
        all_weights['w2'] = tf.Variable(tf.zeros([self.num_hidden,
                        self.num_input],
                        dtype = tf.float32))
        all_weights['b2'] = tf.Variable(tf.zeros([self.num_input],
            dtype = tf.float32))
        return all_weights

    def partial_fit(self, X):
        cost, opt = self.session.run((self.cost, self.optimizer),
                    feed_dict = {self.x: X,
                                self.scale: self.training_scale})
        return cost

    def kl_divergence(self, p, p_hat):
        return tf.reduce_mean(p*(tf.log(p)/tf.log(p_hat)) +
            (1-p)*(tf.log(1-p)/tf.log(1-p_hat)))

    def calculate_total_cost(self, X):
        return self.session.run(self.cost, feed_dict = {
            self.x: X,
            self.scale: self.training_scale
        })

    def transform(self, X):
        return self.session.run(self.hidden_layer,
            feed_dict = {self.x: X, self.scale: self.training_scale})

    def generate(self, hidden = None):
```

```python
    if hidden is None:
        hidden = np.random.normal(size = self.weights["b1"])
    return self.session.run(self.reconstruction,
        feed_dict = {self.hidden_layer: hidden})

def reconstruct(self, X):
    return self.session.run(self.reconstruction,
        feed_dict = {self.x: X, self.scale: self.training_scale})
def get_weights(self):
    return self.session.run(self.weights['w1'])

def get_biases(self):
    return self.session.run(self.weights['b1'])
```

在下一节中，我们将看看应用于特定数据集的稀疏自编码器。

8.6.3 应用于 MNIST 数据集的稀疏自编码器

在和其他示例中相同的数据集上运行该编码器，并比较结果：

```python
class SparseAutoEncoderExample:
    def main(self):
        mnist = input_data.read_data_sets('MNIST_data', one_hot = True)

        def get_random_block_from_data(data, batch_size):
            start_index = np.random.randint(0, len(data) - batch_size)
            return data[start_index:(start_index + batch_size)]

        X_train = mnist.train.images
        X_test = mnist.test.images

        n_samples = int(mnist.train.num_examples)
        training_epochs = 5
        batch_size = 128
        display_step = 1

        autoencoder =SparseAutoencoder(num_input=784,
                                    num_hidden = 200,
                                    transfer_function = tf.nn.sigmoid,
                                    optimizer = tf.train.AdamOptimizer(
                                        learning_rate = 0.001),
                                    scale = 0.01)

        for epoch in range(training_epochs):
            avg_cost = 0.
            total_batch = int(n_samples / batch_size)
            # Loop over all batches
            for i in range(total_batch):
                batch_xs = get_random_block_from_data(X_train, batch_size)
```

```python
            # Fit training using batch data
            cost = autoencoder.partial_fit(batch_xs)
            # Compute average loss
            avg_cost += cost / n_samples * batch_size

        # Display logs per epoch step
        if epoch % display_step == 0:
            print("Epoch:", '%04d' % (epoch + 1), "cost=", avg_cost)

    print("Total cost: " +
        str(autoencoder.calculate_total_cost(X_test)))

    # input weights
    wts = autoencoder.get_weights()
    dim = math.ceil(math.sqrt(autoencoder.num_hidden))
    plt.figure(1, figsize=(dim, dim))
    for i in range(0, autoencoder.num_hidden):
        im = wts.flatten()[i::autoencoder.num_hidden].reshape((28, 28))
        ax = plt.subplot(dim, dim, i + 1)
        for label in (ax.get_xticklabels() + ax.get_yticklabels()):
            label.set_fontsize(6)
        plt.subplot(dim, dim, i+1)
        plt.imshow(im, cmap="gray", clim=(-1.0, 1.0))
    plt.suptitle('Sparse AutoEncoder Weights', fontsize=15, y=0.95)
    plt.savefig('figures/sparse_autoencoder_weights.png')
    plt.show()

    predicted_imgs = autoencoder.reconstruct(X_test[:100])

    # plot the reconstructed images
    plt.figure(1, figsize=(10, 10))
    plt.title('Sparse Autoencoded Images')
    for i in range(0,100):
        im = predicted_imgs[i].reshape((28,28))
        ax = plt.subplot(10, 10, i + 1)
        for label in (ax.get_xticklabels() + ax.get_yticklabels()):
            label.set_fontsize(6)

        plt.subplot(10, 10, i+1)
        plt.imshow(im, cmap="gray", clim=(0.0, 1.0))
    plt.suptitle('Sparse AutoEncoder Images', fontsize=15, y=0.95)
    plt.savefig('figures/sparse_autoencoder_images.png')
    plt.show()

def main():
    auto = SparseAutoEncoderExample()
    auto.main()

if __name__ == '__main__':
    main()
```

代码的输出如下所示：

```
('Epoch:', '0001', 'cost=', 1697.039439488638)
('Epoch:', '0002', 'cost=', 667.23002088068188)
('Epoch:', '0003', 'cost=', 450.02947024147767)
('Epoch:', '0004', 'cost=', 351.54360497159115)
('Epoch:', '0005', 'cost=', 293.73473448153396)
Total cost: 21025.2
```

可以看到损失比其他的编码器低，因此 KL 散度和稀疏性是有用的。

8.6.4 比较稀疏自编码器和加性高斯噪声自编码器

图 8-10 展示了加性高斯噪声自编码器和稀疏自编码器的损失值的比较结果。

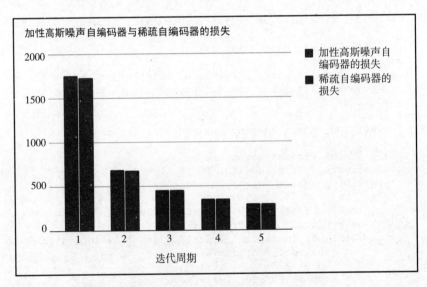

图 8-10　在 MNIST 数据集上 5 个迭代周期后的两种自编码器损失值的比较

8.7 总结

在本章中，你学习了三种不同类型的自编码器：基本自编码器、加性高斯噪声自编码器和稀疏自编码器。我们了解了它们在哪些情况下有用。我们在 MNIST 数据集上训练三种自编码器，并比较了它们之间的损失（情况）。我们还绘制了它们的权重和近似输出的图片。

CHAPTER 9
第 9 章

神经网络研究

在本章中,我们将看一下神经网络中活跃的研究领域。以下分析的问题涉及多个领域,从基础的研究领域到复杂的实践应用。

- 神经网络中的过拟合。
- 使用神经网络进行大规模视频处理。
- 使用双分支互向(twisted)神经网络进行命名实体识别。
- 双向递归神经网络。

9.1 神经网络中避免过拟合

让我们了解过拟合的形成原因以及如何在神经网络中避免过拟合。Nitesh Srivastava、Geoffrey Hinton 等人于 2014 年发表了一篇论文,链接见 https://www.cs.toronto.edu/~hinton/absps/JMLRdropout.pdf,阐述了如何避免过拟合的示例。

9.1.1 过拟合问题阐述

深度神经网络包括非线性隐藏层,这使得它们可以学习输入和输出之间非常复杂的关系,使模型更具有表征性。然而,这些复杂的关系也是导致采样噪声的结果。测试数据中可能并不存在这样复杂的关系，从而导致过拟合。已经有很多技术和方法来消除这种噪声。这些方法包括模型效果在测试集上变差时立即停止训练，引入诸如 L1、L2 正则化的权重惩罚项，软权重共享（Nowlan and Hinton，1992）。

9.1.2 过拟合解决方案

Dropout 是解决模型效果的一项技术。还有一些其他技术诸如多模型平均，它能防止过拟合，并提供了一种能有效综合多种神经网络架构的方法。Dropout 指在神经网络中丢弃神经单元（隐藏层和可见层神经单元）。丢弃一个神经单元，意味着从输入和输出的网络连接中移除，如图 9-1 所示。

丢弃单元的选择通常是随机的。简单来说，每个单元都有一个独立于其他单元的概率 p，p 的选择可以基于验证集，也可以设置为 0.5，这个值对很多神经网络和任务来说都是较好的。

然而，对于输入单元而言，最佳的保留概率通常接近于 1 而不是 0.5。

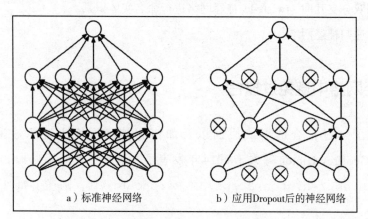

图 9-1　在神经网络中应用 Dropout 技术图示

Dropout 神经网络模型：

❑ 一个具有两个隐藏层的标准神经网络。

❑ 通过对图 9-1a 中的神经网络使用 Dropout 产生一个较窄的神经网络，一些交叉单元已经被丢弃。

以下代码展示了如何在 TensorFlow 中使用 Dropout。

```
cell = tf.nn.rnn_cell.LSTMCell(state_size, state_is_tuple=True)
cell = tf.nn.rnn_cell.DropoutWrapper(cell, output_keep_prob=0.5)
cell = tf.nn.rnn_cell.MultiRNNCell([cell] * num_layers,
state_is_tuple=True)
```

从上面的代码示例中可以看到，对 `LSTMCell` 使用 Dropout 值为 0.5。在这里 `output_keep_prob` 为单位张量或浮点数，取值在 0 到 1 之间。如果取值为常量 1，则表示不使用 Dropout。

9.1.3 影响效果

图 9-2 展示了 Dropout 如何影响模型的准确度。

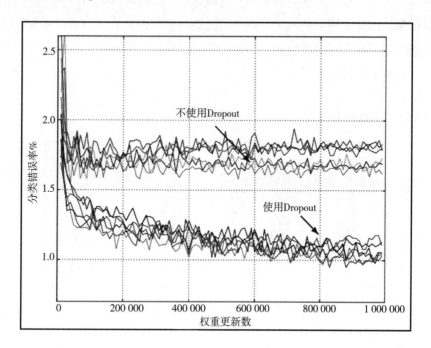

图 9-2　Dropout 对模型的准确度影响图示

如图 9-2 所示，模型使用 Dropout 使得分类错误率有明显下降。

9.2 使用神经网络进行大规模视频处理

在论文（https://static.googleusercontent.com/media/research.google.com/en//pubs/archive/42455.pdf）中，作者探讨了 CNN 如何用于大规模视频分类。在这个实例中，神经网络不仅可以捕获单个静态图像的外观信息，还可以捕获图像随时间的演化信息。在大规模视频场景下应用 CNN 有几个挑战。

与现有图像数据集相匹配的视频分类标准很少（或没有）。因为视频收集、标注、存储有极大的挑战。为获取足够的数据来训练我们的 CNN 网络，作者收集了 Sports-1M 数据集，该数据集来源于 YouTube，数据集包含 100 万个视频，每个视频隶属于 487 个体育类别中的一个。Sports-1M 也可供研究机构用于该领域的研究。

在该数据集中，作者将每个视频视为一组简短的、固定大小的剪辑。每个剪辑包含几个连续的帧，因此，网络的链接可以在时间维度上延伸，并学习到时间和空间上的特征。作者描述了三种宽泛链接模式（前期融合、后期融合和缓慢融合）。之后，我们将考虑多分辨率来解决计算效率问题。

图 9-3 展示了各种融合技术。

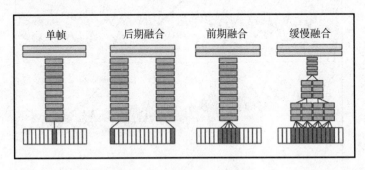

图 9-3　各种融合技术图示

9.2.1 分辨率改进方案

作者使用了一种多分辨率体系结构，旨在通过两个独立的处理流（称为 Fovea 和 Context 流）在两个空间分辨率上达成折衷方案（见图 9-4）。网络的输入是大小为

178×178 的视频剪辑。

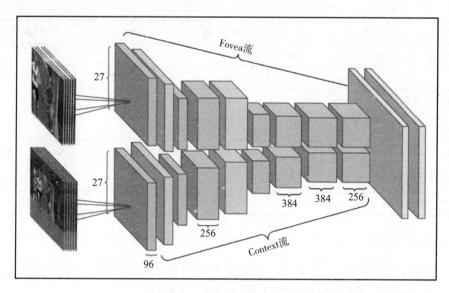

图 9-4 多分辨率 CNN

Context 流的输入是原始空间分辨率的降采样一半（89×89 像素）的结果。Fovea 流的输入是原始分辨率中心的 89×89 区域。这样，总的输入维度就减半了。

9.2.2 特征直方图基线

除了比较 CNN 体系结构之外，作者还提到了一种基于特征的方法的准确性。使用标准 bag-of-words 处理流程在视频的所有帧处提取几种类型的特征，然后使用 k-means 向量量化将它们离散化并利用空间金字塔编码和软量化将词积累成直方图。

9.2.3 定量结果

表 9-1 汇总了 Sports-1M 数据集测试集的结果（200 000 个视频和 4 000 000 个剪辑）。多网络融合的方法始终明显优于基于特征方法的基线。基于特征的方法在视频的持续时间内密集地计算视觉词并产生基于完整视频层面的特征向量预测，而作者提出的网络仅分别查看 20 个随机采样的剪辑。

表 9-1　关于 Sports-1M 测试集的 200 000 个视频的结果。Hit@k 值表示在前 k 个预测中包含至少一个真实标签的测试样本的比例

模　型	剪辑 Hit@1	视频 Hit@1	视频 Hit@5
特征直方图 + 神经网络	—	55.3	—
单帧	41.1	59.3	77.7
单帧 + 多分辨率	**42.4**	**60.0**	**78.5**
单帧仅 Fovea	30.0	49.9	72.8
单帧仅 Context	38.1	56.0	77.2
Early Fusion	38.9	57.7	76.8
Late Fusion	40.7	59.3	78.7
Slow Fusion	**41.9**	**60.9**	**80.2**
CNN 平均值（Single+Early+Late+Slow）	41.4	63.9	82.4

尽管标签有噪声，采用网络拓扑的方法仍然学习得很好；训练数据中的视频受到一些不正确的标注的影响，甚至正确标记的视频通常也会包含大量的人工痕迹，如文字、效果、剪辑和图标，我们都没有试图明确地过滤掉这些痕迹。

9.3　使用双分支互向神经网络进行命名实体识别

在这篇文章（http://www.cs.cmu.edu/~leili/pubs/lu-baylearn2015-twinet.pdf）中，作者着眼于自然语言中实体识别的问题。这通常是问答、对话和其他许多 NLP 用例的第一步。对于文本标记序列问题，命名实体识别会标记出属于个人和组织机构类别的词块。

9.3.1　命名实体识别的例子

IOB 标记系统是 NER 的惯例之一。IOB 标记系统包含以下形式的标签：

- B-{CHUNK_TYPE}：对于在 CHUNK_TYPE 块**开始**的词。
- I-{CHUNK_TYPE}：对于在 CHUNK_TYPE 块**中**的词。
- O：在任何块之外。

- B-PERSON：人物实体。
- B-GPE：地缘政治实体。

以下文本显示了一个句子中命名实体的示例：

```
John      has  lived  in  Britain   for  14         years      .
B-PERSON  O    O      O   B-GPE     O    B-CARDINAL I-CARDINAL O
```

然而，因为以下 2 个原因，这是相当具有挑战性的：

- 实体数据库通常是不完整的（因为有很多新建立的组织）。
- 根据上下文的不同，同一个短语可以指代不同的实体（或无实体）。

9.3.2　定义 Twinet

双分支互向 RNN（Twinet）使用两个并行分支。每个分支由一个递归网络层、一个非线性感知器层和一个反向递归网络层组成。两个分支相互扭在一起：在第二个分支中层的顺序相反。所有递归网络层的输出都会被收集起来。

回顾一下，**递归神经网络（RNN）**的输入是一系列输入向量 $x1..T$，然后递归地计算隐藏状态（也称为输出标签）：

$$h_t = \sigma(U \cdot x_t + W \cdot h_t-1)$$

其中

- t 是 $1..T$。
- x_t 是 t 时刻的外部输入。
- W 是权重。
- h_{t-1} 是 $t-1$ 时刻隐藏层的权重。
- h_t 是 t 时刻正在计算的权重。
- U 是 tanh 层的权重，它参与计算 t 时刻隐藏层的权重。

$\sigma(\cdot)$ 是非线性激活函数。在作者做的实验中,使用**整流线性函数(RELU)**。

9.3.3 结果

在与斯坦福 NER 和伊利诺伊州 NER 的比较中可以看到,Twinet 有明显优势(见图 9-5)。NER 识别器序号用 60 表示(命名实体识别器)。

	P	R	F1
斯坦福NER	84.04	80.96	82.42
伊利诺伊州NER	85.86	84.20	85.02
Twinet	86.06	8634	**86.20**

图 9-5 比较示意图

从图 9-5 中可以看出,在精确度(P)、召回率(R)以及 F1 的得分方面,Twinet 都更高。

9.4 双向递归神经网络

在本节中,我们将关注一个在 NLP 领域正在崛起的新型神经网络拓扑结构。

Schuster 和 Paliwal 在 1997 年引入了**双向递归神经网络(BRNN)**。BRNN 增加了网络可以获取的输入信息量。我们知道**多层感知器(MLP)**和时间延迟神经网络(TDNN)对输入数据灵活性具有限制。RNN 也需要修改其输入数据。更高级的网络拓扑(如 RNN)对输入数据也有限制,因为未来的输入信息无法从当前状态预测。相反,BRNN 不需要修改其输入数据。其未来的输入信息可以从当前状态中获得。BRNN 的想法是将两个相反方向的隐藏层连接到相同的输出。通过这种结构,输出层能够获取过去和未来状态的信息。

当需要输入的上下文时，BRNN 很有用。例如，在手写识别中，可以通过了解当前字母之前和之后的字母来提高性能（见图 9-6）。

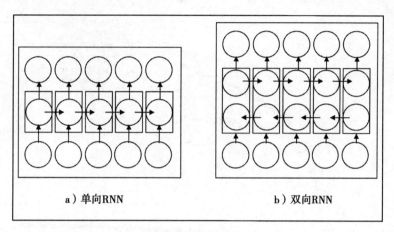

a）单向RNN　　　　　　　b）双向RNN

图 9-6　结构视图

在 TIMIT 数据集中应用 BRNN

在本节中，我们将看看如何在 TIMIT 数据集中用 BRNN 提高音素文本分类的准确率。

TIMIT 是一个由不同性别和方言的美国人的所录制而成的英语音素和词汇级连续语音语料库。每个录制的元素都会被及时记录下来。TIMIT 旨在进一步提高声学语音知识和自动语音识别系统的性能。

如图 9-7 所示，与 MLP 相比，BRNN 无论是在训练集还是测试集上的准确率都更高。而 BLSTM 具有更高的准确率。

9.5　总结

在本章中，我们介绍了一些已经完成研究的领域。这其中包括提升准确率，避免过拟合。我们也看了一些相对较新的领域，比如视频分类。在本书的范围之外，除了 ACM 和 IEEE 一级会议之外，我们诚恳地建议你去 Google、FaceBook、百度等网站搜索正在进行的研究。

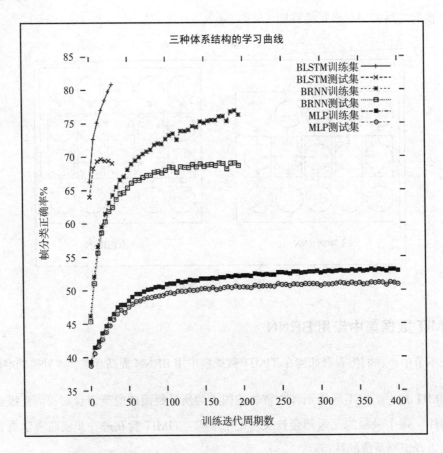

图 9-7　MLP、BRNN 和 BLSTM 的学习曲线对比图

CHAPTER 10
第 10 章

开始使用 TensorFlow

TensorFlow 是 Google 开源的深度学习库。它提供了用张量定义函数并自动计算它们导数的单元。张量可以表示为一个多维数组。标量、向量和矩阵是张量的类型。我们主要用 TensorFlow 设计计算图，构建和训练深度学习模型。TensorFlow 库使用数据流图进行数值计算，其中节点表示数学运算，边表示数据点（通常在这些边之间传输多维数组或张量）。

10.1 环境搭建

最好使用 PyCharm 等 IDE 来编辑 Python 代码；它提供了更快的开发工具和编码助手。代码补全和代码检查使代码编写和调试更快更简单，这可以让你专注于编写神经网络代码。

TensorFlow 提供了多种语言的 API：Python、C++、Java、Go 等。下载一个让我们能够使用 Python 编写深度学习模型的 TensorFlow 版本。在 TensorFlow 安装网站上，可以找到用 virtualenv、pip 和 Docker 安装 TensorFlow 的常用方式和最新说明。

以下步骤描述如何搭建本地开发环境：

1）下载 Pycharm 社区版。

2）在 Pycharm 上获取最新的 Python 版本。

3）转到 Preferences 选项，设置 Python 解释器，并安装最新版本的 TensorFlow：

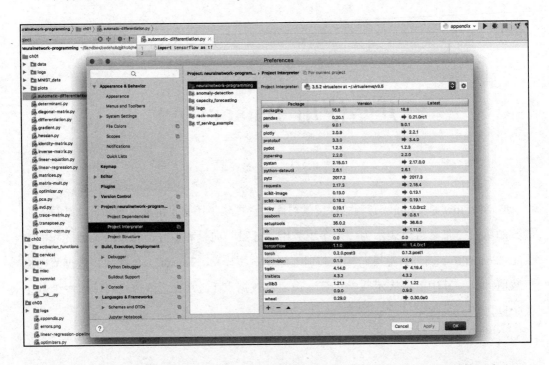

4）TensorFlow 现在将出现在已安装的软件包列表中。点击 OK。现在使用如 hello world 等程序测试安装是否成功：

```
import TensorFlow as tf
helloWorld = tf.constant("Hello World!")
sess = tf.Session()
print(sess.run(helloWorld))
```

10.2 比较 TensorFlow 和 Numpy

TensorFlow 和 Numpy 都是 N 维数组库。但 TensorFlow 还允许我们创建张量函数并计算导数。TensorFlow 已经成为深度学习的主要库之一，因为它非常高效且可以在 GPU 上运行。

以下程序描述了如何用 TensorFlow 和 numpy 执行类似创建（3,3）形状的张量的操作：

```
import TensorFlow as tf
import numpy as np

tf.InteractiveSession()

# TensorFlow operations
a = tf.zeros((3,3))
b = tf.ones((3,3))

print(tf.reduce_sum(b, reduction_indices=1).eval())
print(a.get_shape())

# numpy operations
a = np.zeros((3, 3))
b = np.ones((3, 3))
print(np.sum(b, axis=1))
print(a.shape)
```

输出结果如下：

```
[ 3.  3.  3.]
(3, 3)
[ 3.  3.  3.]
(3, 3)
```

10.3 计算图

TensorFlow 是基于计算图的深度学习库。计算图是一个包含很多节点的网络，其中每个节点定义一个运行函数的操作；这可以像加法或减法一样简单，也可以像多变量方程那样复杂。TensorFlow 程序的构建阶段是在生成计算图的时候，程序的执行阶段是在利用会话对象执行图中的操作的时候。

操作称为 op，可以返回零个或多个张量，我们可以稍后在图中使用这些张量。我们可以给每个操作一个常量、数组或 n 维矩阵。

10.3.1 图

当导入 TensorFlow 库时，默认的图会被实例化。在一个文件中创建不相互依赖的多个模型时，构建一个图对象而不使用默认图会非常有用。常量和操作会被添加到 TensorFlow 的图中。

在 `newGraph.as_default()` 之外应用的变量和操作将被添加到导入库时创建的默认图中：

```
newGraph = tf.Graph()
with newGraph.as_default():
    newGraphConst = tf.constant([2., 3.])
```

10.3.2 会话对象

TensorFlow 中的会话封装了评估张量对象的环境。会话可以拥有指定的私有变量、队列和读取器。我们应该在会议结束时使用 `close` 方法。

会话有 3 个可选参数：

- `Target`：连接到的执行引擎。
- `graph`：要启动的图对象。
- `config`：这是一个 ConfigProto 的协议缓冲区。

如果要单步执行 TensorFlow 的计算，将调用阶跃函数，并执行图的必要依赖：

```
# session objects
a = tf.constant(6.0)
b = tf.constant(7.0)

c = a * b
with tf.Session() as sess:
    print(sess.run(c))
    print(c.eval())
```

`sess.run(c)` 在当前激活的会话中！

代码的执行结果如下：

```
42.0, 42.0
```

`tf.InteractiveSession()` 函数是在 `ipython` 中打开默认会话的简单方法。

sess.run(c) 是一个 TensorFlow Fetch 的例子：

```
session = tf.InteractiveSession()
cons1 = tf.constant(1)
cons2 = tf.constant(2)
cons3 = cons1 + cons2
# instead of sess.run(cons3)
cons3.eval()
```

10.3.3 变量

在训练模型时，我们使用变量来保存和更新参数。变量就像包含张量的内存缓冲区。我们以前使用的所有张量都是常量张量，而不是变量。

变量由会话对象管理或维护。变量在会话之间一直存在，这很有用，因为张量和操作对象是不可变的：

```
# tensor variables
 W1 = tf.ones((3,3))
W2 = tf.Variable(tf.zeros((3,3)), name="weights")

 with tf.Session() as sess:
    print(sess.run(W1))
    sess.run(tf.global_variables_initializer())
    print(sess.run(W2))
```

代码的输出如下：

```
[[ 1.  1.  1.] [ 1.  1.  1.] [ 1.  1.  1.]]
[[ 0.  0.  0.] [ 0.  0.  0.] [ 0.  0.  0.]]
```

TensorFlow 变量在赋值之前必须先初始化，这和常量张量不同：

```
# Variable objects can be initialized from constants or random values
W = tf.Variable(tf.zeros((2,2)), name="weights")
R = tf.Variable(tf.random_normal((2,2)), name="random_weights")

with tf.Session() as sess:
   # Initializes all variables with specified values.
   sess.run(tf.initialize_all_variables())
   print(sess.run(W))
   print(sess.run(R))
```

代码的输出如下：

```
[[ 0.   0.] [ 0.   0.]]
[[ 0.65469146 -0.97390586] [-2.39198709  0.76642162]]

state = tf.Variable(0, name="counter")
new_value = tf.add(state, tf.constant(1))
update = tf.assign(state, new_value)

with tf.Session() as sess:
    sess.run(tf.initialize_all_variables())
    print(sess.run(state))
    for _ in range(3):
        sess.run(update)
        print(sess.run(state))
```

输出如下：

```
0 1 2 3
```

获取变量的状态：

```
input1 = tf.constant(5.0)
input2 = tf.constant(6.0)
input3 = tf.constant(7.0)
intermed = tf.add(input2, input3)
mul = tf.multiply(input1, intermed)

# Calling sess.run(var) on a tf.Session() object retrieves its value. Can
retrieve multiple variables simultaneously with sess.run([var1, var2])
with tf.Session() as sess:
    result = sess.run([mul, intermed])
    print(result)
```

输出结果如下：

```
[65.0, 13.0]
```

10.3.4 域

TensorFlow 模型可能有数百个变量。`tf.variable_scope()` 可以为变量提供简单的名字。

为了管理复杂模型并将其分解成单独的部分，TensorFlow 具有域的概念。域非常简单，当使用 TensorBoard 时，域很有用。域也可以嵌套在其他域内：

```
with tf.variable_scope("foo"):
    with tf.variable_scope("bar"):
        v = tf.get_variable("v", [1])
assert v.name == "foo/bar/v:0"

with tf.variable_scope("foo"):
    v = tf.get_variable("v", [1])
    tf.get_variable_scope().reuse_variables()
    v1 = tf.get_variable("v", [1])
assert v1 == v
```

以下示例显示如何使用 reuse 选项来理解 `get_variable` 的工作原理：

```
#reuse is false
with tf.variable_scope("foo"):
    n = tf.get_variable("n", [1])
assert v.name == "foo/n:0"

#Reuse is true
with tf.variable_scope("foo"):
    n = tf.get_variable("n", [1])
with tf.variable_scope("foo", reuse=True):
    v1 = tf.get_variable("n", [1])
assert v1 == n
```

10.3.5 数据输入

外部数据导为 TensorFlow 对象：

```
a = np.zeros((3,3))
ta = tf.convert_to_tensor(a)
with tf.Session() as sess:
    print(sess.run(ta))
```

输出结果如下：

`[[ 0. 0. 0.] [ 0. 0. 0.] [ 0. 0. 0.]]`

10.3.6 占位符和输入字典

使用 `tf.convert_to_tensor()` 可以很方便地输入数据，但不能对数据进行缩放。使用 `tf.placeholder` 变量（为计算图提供数据入口的虚拟节点）。`feed_dict` 是一个 Python 字典映射：

```
input1 = tf.placeholder(tf.float32)
 input2 = tf.placeholder(tf.float32)
 output = tf.multiply(input1, input2)

with tf.Session() as sess:
    print(sess.run([output], feed_dict={input1:[5.], input2:[6.]}))
```

代码的输出如下:

```
[array([ 30.], dtype=float32)]
```

10.4 自动微分

自动微分也称为**算法微分**,它是自动数值计算函数导数的方法。它有助于计算数值优化等应用中有用的梯度、Jacobian 和 Hessian。反向传播算法是自动微分的反向模式的实现。

在以下示例中,我们在 `mnsit` 数据集上使用其中一个损失函数计算损失。问题是:我们如何使模型拟合数据呢?

我们可以用 `tf.train.Optimizer` 构建一个优化器。

`tf.train.Optimizer.minimize(loss, var_list)` 向计算图添加优化操作,并自动计算梯度而无需用户输入:

```
import TensorFlow as tf

# get mnist dataset
from TensorFlow .examples.tutorials.mnist import input_data
data = input_data.read_data_sets("MNIST_data/", one_hot=True)

# x represents image with 784 values as columns (28*28), y represents output digit
x = tf.placeholder(tf.float32, [None, 784])
y = tf.placeholder(tf.float32, [None, 10])
# initialize weights and biases [w1,b1][w2,b2]
numNeuronsInDeepLayer = 30
w1 = tf.Variable(tf.truncated_normal([784, numNeuronsInDeepLayer]))
b1 = tf.Variable(tf.truncated_normal([1, numNeuronsInDeepLayer]))
w2 = tf.Variable(tf.truncated_normal([numNeuronsInDeepLayer, 10]))
```

```
b2 = tf.Variable(tf.truncated_normal([1, 10]))

# non-linear sigmoid function at each neuron
def sigmoid(x):
    sigma = tf.div(tf.constant(1.0), tf.add(tf.constant(1.0),
tf.exp(tf.negative(x))))
    return sigma

# starting from first layer with wx+b, then apply sigmoid to add non-
linearity
z1 = tf.add(tf.matmul(x, w1), b1)
a1 = sigmoid(z1)
z2 = tf.add(tf.matmul(a1, w2), b2)
a2 = sigmoid(z2)

# calculate the loss (delta)
loss = tf.subtract(a2, y)

# derivative of the sigmoid function der(sigmoid)=sigmoid*(1-sigmoid)
def sigmaprime(x):
    return tf.multiply(sigmoid(x), tf.subtract(tf.constant(1.0),
sigmoid(x)))

# automatic differentiation
cost = tf.multiply(loss, loss)
step = tf.train.GradientDescentOptimizer(0.1).minimize(cost)

acct_mat = tf.equal(tf.argmax(a2, 1), tf.argmax(y, 1))
acct_res = tf.reduce_sum(tf.cast(acct_mat, tf.float32))

sess = tf.InteractiveSession()
sess.run(tf.global_variables_initializer())

for i in range(10000):
    batch_xs, batch_ys = data.train.next_batch(10)
    sess.run(step, feed_dict={x: batch_xs,
                              y: batch_ys})
    if i % 1000 == 0:
        res = sess.run(acct_res, feed_dict=
        {x: data.test.images[:1000],
         y: data.test.labels[:1000]})
        print(res)
```

10.5 TensorBoard

　　TensorFlow 拥有强大的内置可视化工具 TensorBoard。它可以让开发人员解释、可视化和调试计算图。为了在 TensorBoard 中自动可视化图和度量，TensorFlow 将和执行计算图相关的事件写入特定文件夹。

以下例子显示了之前分析过的计算图:

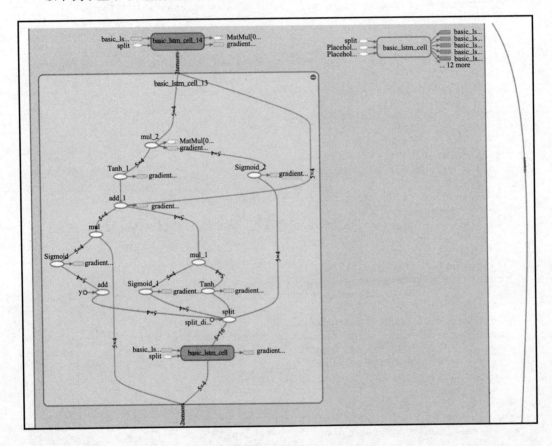

要查看图,请单击 TensorBoard 顶部面板上的图形选项卡。如果图有多个节点,那么在单个视图中可能很难对其进行可视化。为了使我们的可视化更简单,可以使用 tf.name_scope 将具有特定名称的相关操作放到组中。

推荐阅读

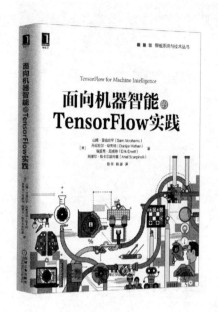 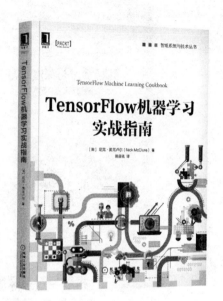

面向机器智能的TensorFlow实践

作者：Sam Abrahams, Danijar Hafner, Erik Erwitt, Dan Becker ISBN：978-7-111-56389-1 定价：69.00元

本书是一本绝佳的TensorFlow入门指南。几位作者都来自谷歌研发一线，他们用自己的宝贵经验，结合众多高质量的代码，生动讲解TensorFlow的底层原理，并从实践角度介绍如何将两种常见模型——深度卷积网络、循环神经网络应用到图像理解和自然语言处理的典型任务中。此外，还介绍了在模型部署和编程中可用的诸多实用技巧。

TensorFlow机器学习实战指南

作者：Nick McClure ISBN：978-7-111-57948-9 定价：69.00元

本书由资深数据科学家撰写，从实战角度系统讲解TensorFlow基本概念及各种应用实践。真实的应用场景和数据，丰富的代码实例，详尽的操作步骤，带你由浅入深系统掌握TensorFlow机器学习算法及其实现。

推荐阅读

机器学习与深度学习：通过C语言模拟

作者：[日]小高知宏 著　译者：申富饶 于僡译　ISBN：978-7-111-59994-4　定价：59.00元

本书以深度学习为关键字讲述机器学习与深度学习的相关知识，对基本理论的讲述通俗易懂，不涉及复杂的数学理论，适用于对机器学习与深度学习感兴趣的初学者。当前机器学习的书籍一般只讲述理论，没有具体的程序实例。有些以实例为主的机器学习书籍则依赖于一些函数库或工具，无法理解其内部算法原理。本书没有使用任何外部函数库或工具，通过C语言程序来实现机器学习和深度学习算法，读者不太理解相关理论时，可以通过C语言程序代码来进行学习。

本书从强化学习、蚁群最优化方法、神经网络、深度学习等出发，分阶段介绍机器学习的各种算法，通过分析C语言程序代码，实际执行C语言程序，使读者能快速步入机器学习和深度学习殿堂。

自然语言处理与深度学习：通过C语言模拟

作者：[日]小高知宏 著　译者：申富饶 于僡译　ISBN：978-7-111-58657-9　定价：49.00元

本书初步探索了将深度学习应用于自然语言处理的方法。概述了自然语言处理的一般概念，通过具体实例说明了如何提取自然语言文本的特征以及如何考虑上下文关系来生成文本。书中自然语言文本的特征提取是通过卷积神经网络来实现的，而根据上下文关系来生成文本则利用了循环神经网络。这两个网络是深度学习领域中常用的基础技术。

本书通过实现C语言程序来具体讲解自然语言处理与深度学习的相关技术。本书给出的程序都能在普通个人电脑上执行。通过实际执行这些C语言程序，确认其运行过程，并根据需要对程序进行修改，能够更深刻地理解自然语言处理与深度学习技术。